Applied SolidWorks

L. Scott Hansen, Ph.D.

Associate Professor of Technology
College of Computing, Integrated Engineering and Technology

Southern Utah University
Cedar City, Utah

Industrial Press, Inc
New York

Industrial Press, Inc.
200 Madison Avenue
New York, NY - 10016-4078

10   9   8   7   6   5   4   3   2   1

Table of Contents

Note to the Reader

This book provides clear and concise applied instruction in order to help you develop a mastery of *SolidWorks*. Almost every instruction includes a graphic illustration to aid in clarifying that instruction. Software commands appear in **bold** or in "quotation marks" for anyone who prefers not to read every word of the text. Most illustrations also include small pointer arrows and text to further clarify instructions.

This book was written for classroom instruction or self-study, including for individuals with no solid modeling experience at all. You will begin at a very basic level, but by the time you finish you will be completing complex functions.

For any organization requiring additional help, I am available for onsite training. Please contact me at shansen1@infowest.com.

Scott Hansen
Cedar City, Utah

Chapter 1 Getting Started

Objectives:

- Create a simple sketch using the Sketch commands
- Dimension a sketch using the Smart Dimension command
- Extrude a sketch using the Extrude command
- Create a hole using the Extrude command
- Create a fillet using the Fillet command
- Create a counter bore using the Hole Wizard

Chapter 1 includes instruction on how to design the part shown below.

1.   Start SolidWorks by moving the cursor to the  button in the lower left corner of the screen. Click the left mouse button once.

2.   A pop up menu of the programs that are installed on the computer will appear. Scroll through the list of programs until you find "SolidWorks".

3.   Move the cursor over the text "SolidWorks" and left click once as shown in Figure 1.

Figure 1

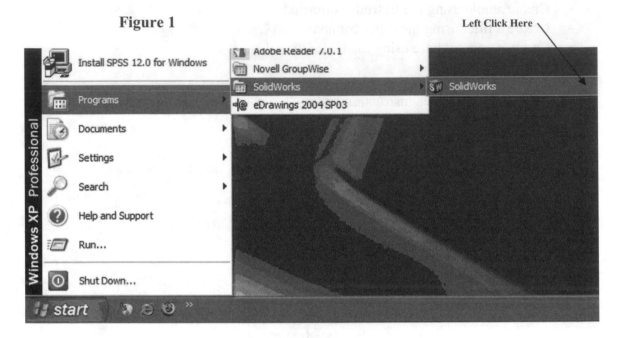

4.   SolidWorks will open (load up and begin running).

5.   Move the cursor to the upper left corner of the screen and left click on the "New Document" icon as shown in Figure 2.

Figure 2   Left Click Here

6.   The New SolidWorks Document dialog box will appear. Left click on **Part** as shown in Figure 3.

Figure 3

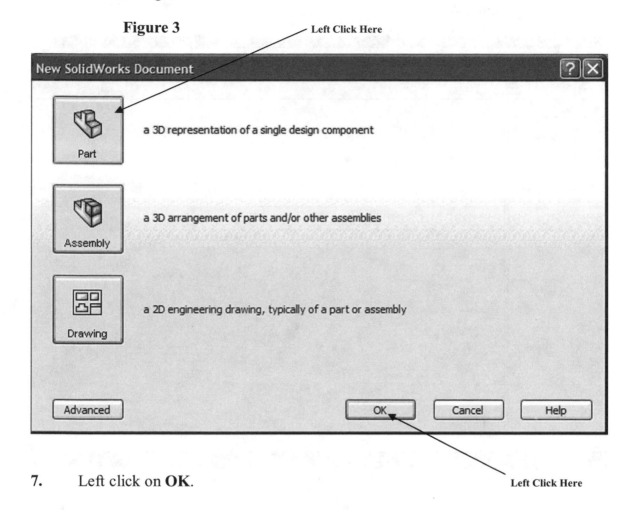

7.   Left click on **OK**.

8.   The screen should look similar to Figure 4.

Figure 4

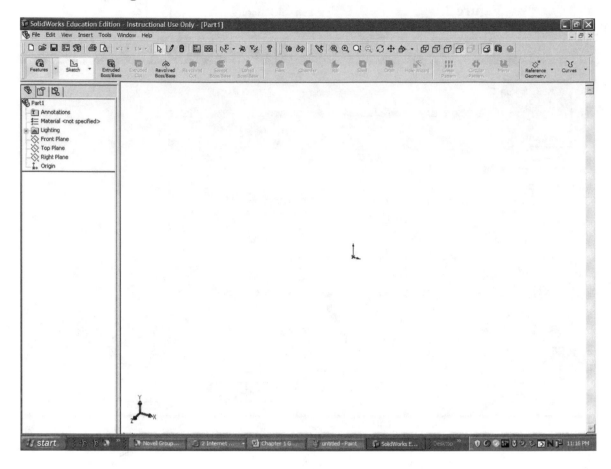

9. Move the cursor to the upper middle portion of the screen and left click on the "Sketch" icon. SolidWorks will display three different work planes. Move the cursor over the "Front Plane" until it turns red. Left click once as shown in Figure 5. SolidWorks will provide a perpendicular view of the front plane.

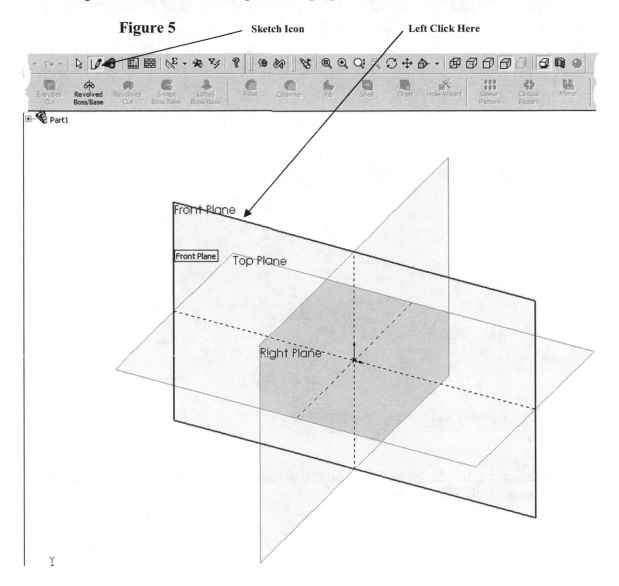

Figure 5

Sketch Icon

Left Click Here

10. Your screen should look similar to Figure 6.

Figure 6

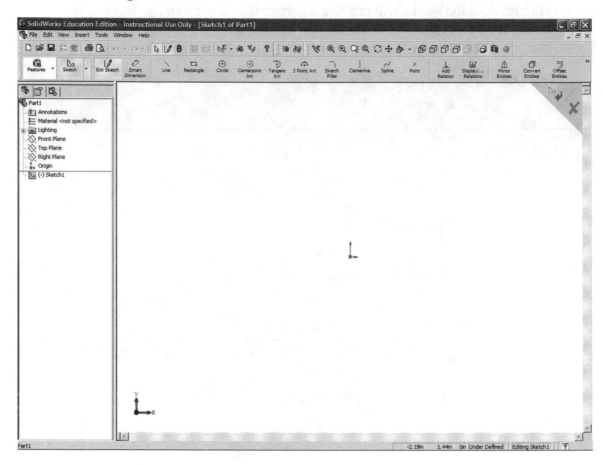

11. If the grid is displayed, skip to instruction number 12. If there is no grid visible, continue following the next steps.

NOTE: Move the cursor to the upper left portion of the screen and left click on **Tools** as shown in Figure A.

Figure A

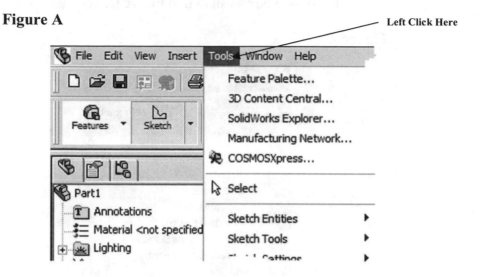

A drop down menu will appear. Left click on **Options** as shown in Figure B.

Figure B

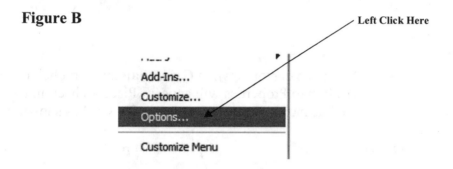

The Document Properties - Detailing dialog box will appear. Left click on the **Document Properties** tab at the upper left portion of the dialog box as shown in Figure C.

Figure C

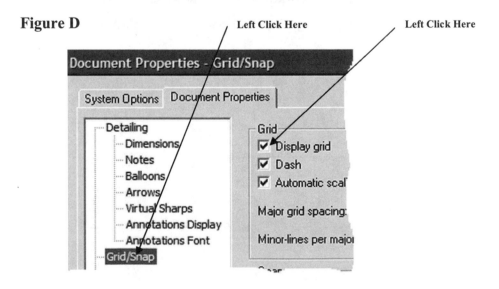

Move the cursor down to **Grid/Snap** and left click once. The Grid/Snap Properties will appear. Place a check in the box (with the left mouse button) next to Display Grid as shown in Figure D.

Figure D

8

Left click on **OK** as shown in Figure E.

Figure E

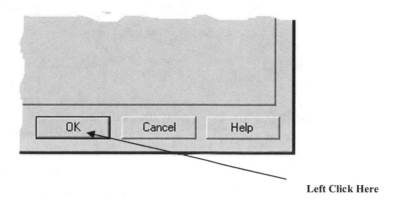

Left Click Here

SolidWorks should now display a visible grid as shown in Figure F.

Figure F

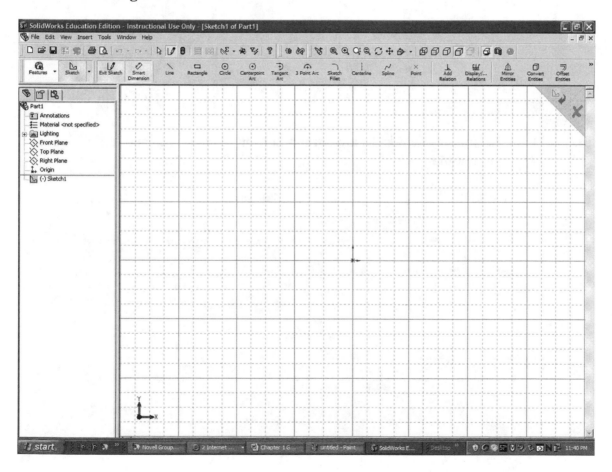

12.   SolidWorks is now ready for use. Left click on **Line** as shown in
Figure 7.

Figure 7

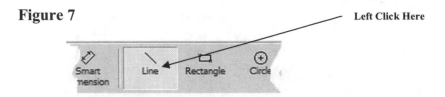

13.   Move the cursor in the lower left portion of the screen and left click once. This
will be the beginning end point of a line as shown in Figure 8.

Figure 8

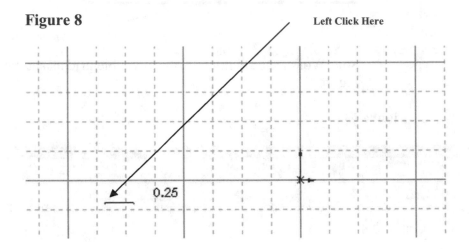

14.   Move the cursor towards the lower right portion of the screen and left click once
as shown in Figure 9.

Figure 9

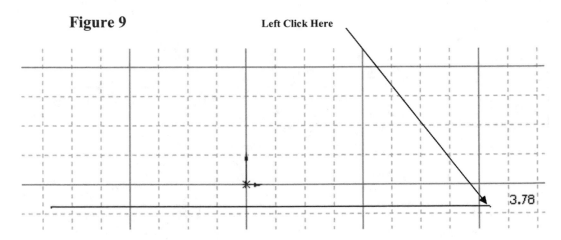

15. While the line is still attached to the cursor, move the cursor towards the top of the screen and left click once. Notice the length of the line is attached to the cursor at the right as shown in Figure 10.

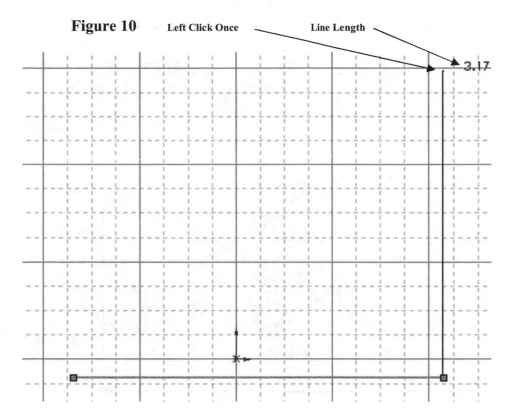

Figure 10   Left Click Once   Line Length

16.   With the line still attached to the cursor, move the cursor towards the left side of the screen. Notice the line of small dots connecting the first and last point together. Left click once when the small dots appear as shown in Figure 11.

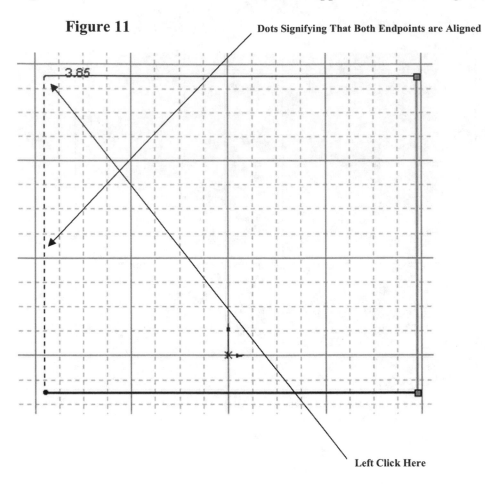

Figure 11

Dots Signifying That Both Endpoints are Aligned

3.85

Left Click Here

17.   Your screen should look similar to Figure 12.

Figure 12

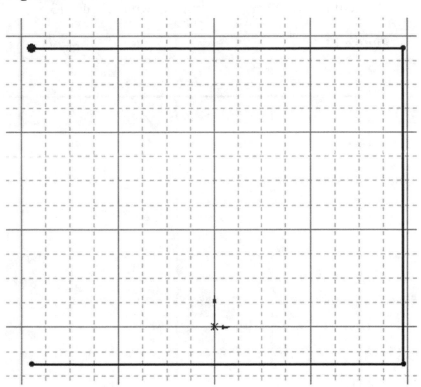

18. Move the cursor down towards the original starting point. Ensure that a red dot appears at the intersection of the two lines. This indicates that SolidWorks has "snapped" to the intersection of the lines. After the red dot appears, left click once. This will form a 90 degree box as shown in Figure 13.

Figure 13

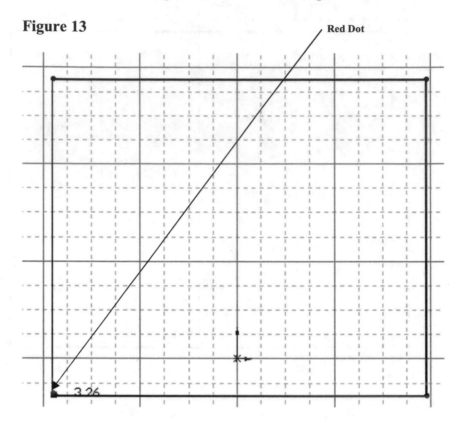

Red Dot

3.26

19. Your screen should look similar to Figure 14.

Figure 14

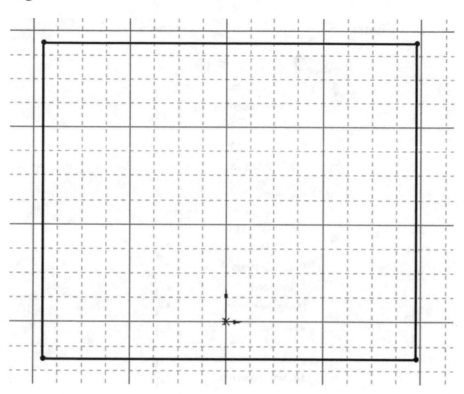

20. Move the cursor to the upper left portion of the screen and left click on **Tools**. A drop down menu will appear. Left click on **Options** located at the bottom of the menu as shown in Figure 15.

Figure 15

Left Click Here

Left Click Here

21. The Document Properties – Detailing dialog box will appear. Left click on the **Document Properties** tab. Left click on **Detailing**. Left click on the drop down arrow below the text "Dimensioning standard". A drop down menu will appear. Left click on **ANSI** as shown in Figure 16.

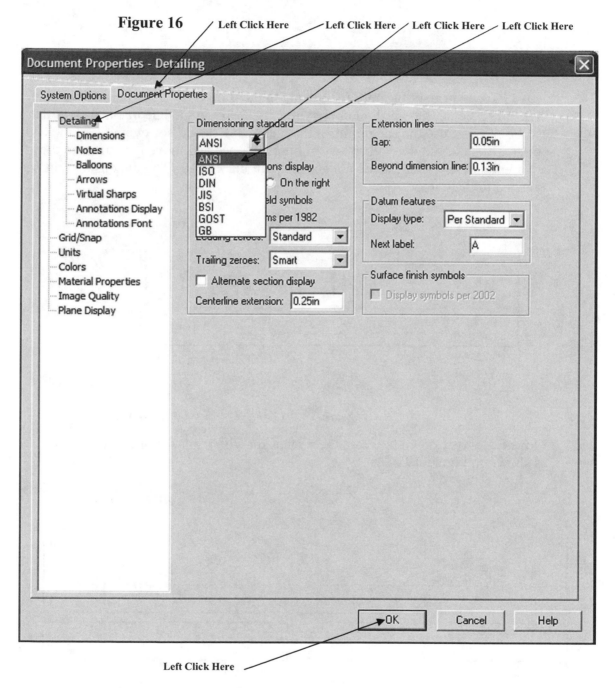

Figure 16

22. Left click on **OK**.

23. Move the cursor to the upper left portion of the screen and left click on **Smart Dimension** as shown in Figure 17.

Figure 17

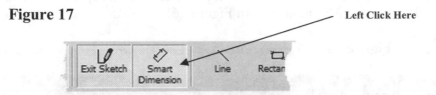

Left Click Here

24. Move the cursor over the bottom horizontal line until it turns red as shown in Figure 18. Select the line by left clicking anywhere on the line **or** on each of the end points. To use the end points of the line, move the cursor over one of the end points. A small red dot will appear. Left click once and move the cursor to the other end point. After the red dot appears, left click once. The dimension will be attached to the cursor.

Figure 18

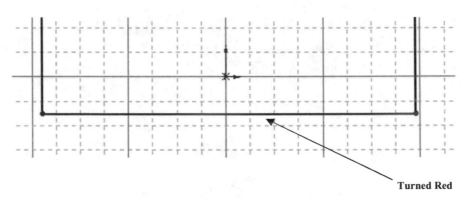

Turned Red

25. Move the cursor down to where the dimension will be placed and left click once as shown in Figure 19.

Figure 19

Left Click Here

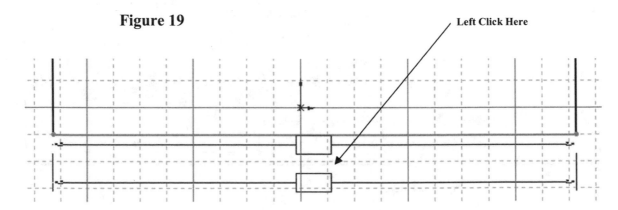

26. The Modify dialog box will appear as shown in Figure 20.

Figure 20

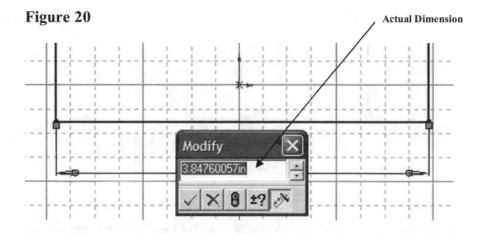

27. To edit the dimension, type **2.00** in the Modify dialog box (while the current dimension is highlighted) and left click on the green checkmark as shown in Figure 21.

Figure 21

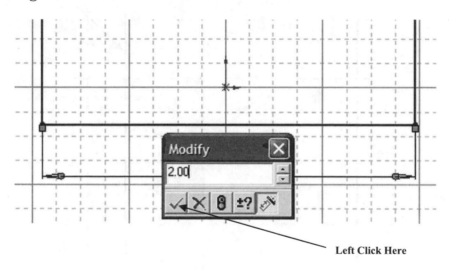

28. The dimension of the line will become 2.00 inches as shown in Figure 22.

Figure 22

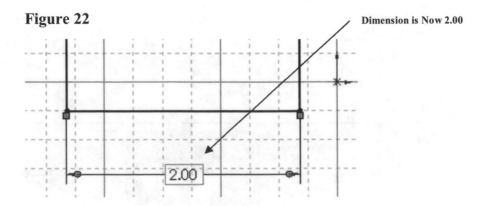

Dimension is Now 2.00

29. Move the cursor over the right side vertical line until it turns red as shown in Figure 23. Left click once.

Figure 23

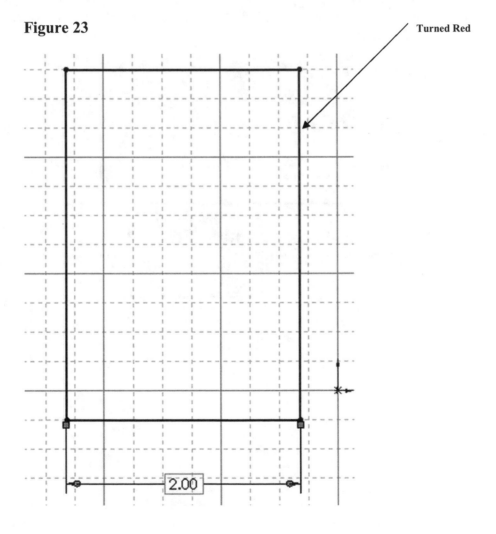

Turned Red

30.   Move the cursor to where the dimension will be placed and left click once as shown in Figure 24.

Figure 24

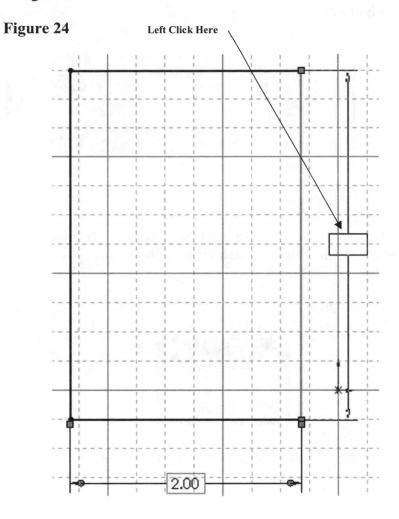

31. The Modify dialog box will appear as shown in Figure 25.

Figure 25

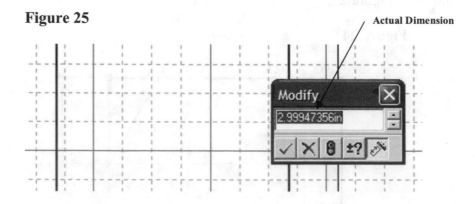

32. To edit the dimension, type **.25** in the Modify dialog box (while the current dimension is highlighted) and left click on the green checkmark as shown in Figure 26.

Figure 26

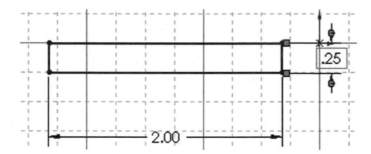

33. The screen should look similar to Figure 27.

Figure 27

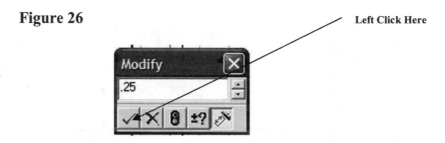

34. Right click anywhere around the drawing. A pop up menu will appear. Left click on **Select** as shown in Figure 28. This verifies that no commands are active.

Figure 28

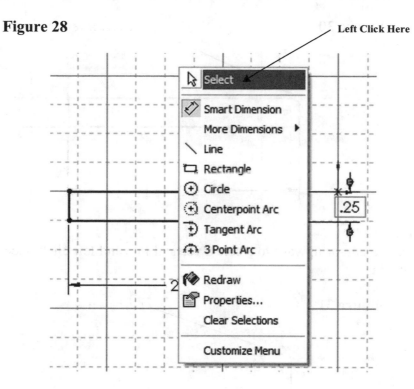

35. Move the cursor to the upper left corner of the screen and left click on **Line** as shown in Figure 29.

Figure 29

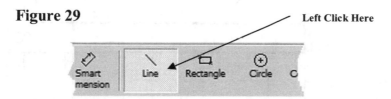

36. Move the cursor to the upper left corner of the box as shown in Figure 30 and left click once.

Figure 30

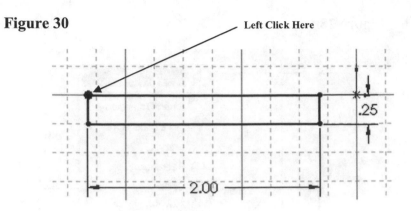

37. Move the cursor upward to create a vertical line and left click once as shown in Figure 31.

Figure 31

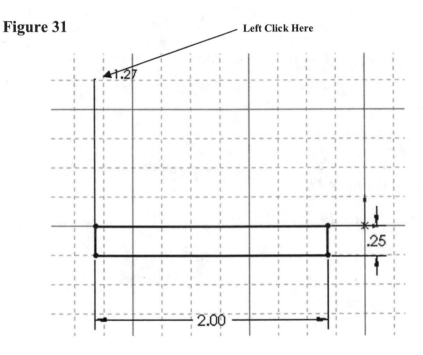

38. Right click the mouse. A pop up menu will appear. Left click on **Select** as shown in Figure 32. This verifies that no commands are active.

Figure 32

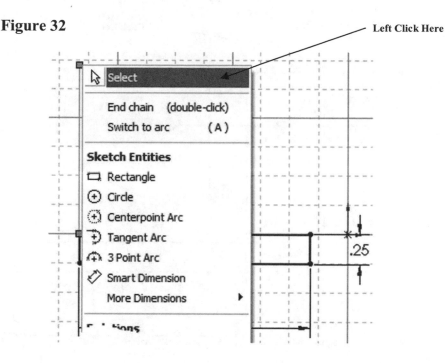

39. You may have to zoom out to leave enough room to construct the following lines. To zoom out use the scroll wheel on the mouse or move the cursor to the upper middle portion of the screen and left click on the "Zoom In/Out" icon as shown in Figure 33.

Figure 33

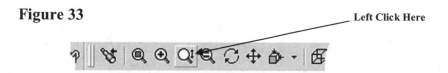

40. After left clicking on the zoom in/out icon, the cursor will change to a magnifying glass. Move the cursor to the middle of the screen. Hold the left mouse button down and drag the cursor up to zoom in or down to zoom out.

41. Move the cursor to the upper left corner of the screen and left click on **Line** as shown in Figure 34.

Figure 34

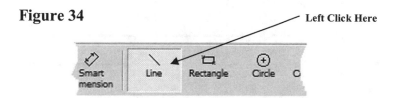

42. Left click on the upper endpoint of the line. Move the cursor to the left and left click once as shown in Figure 35.

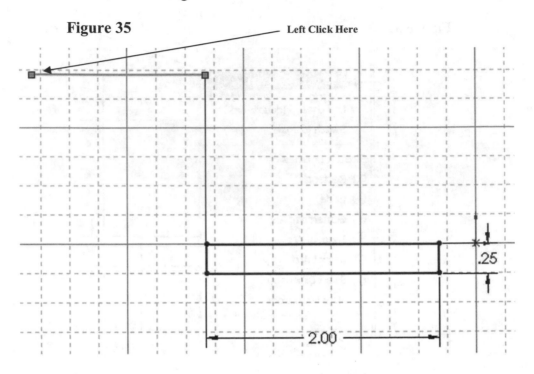

Figure 35

Left Click Here

.25

2.00

43. With the line still attached to the cursor, move the cursor up and left click once as shown in Figure 36.

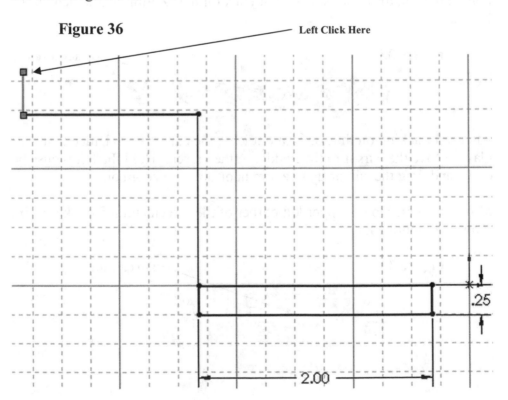

Figure 36

Left Click Here

.25

2.00

44.   With the line still attached to the cursor, move the cursor to the right side of the screen and left click once as shown in Figure 37.

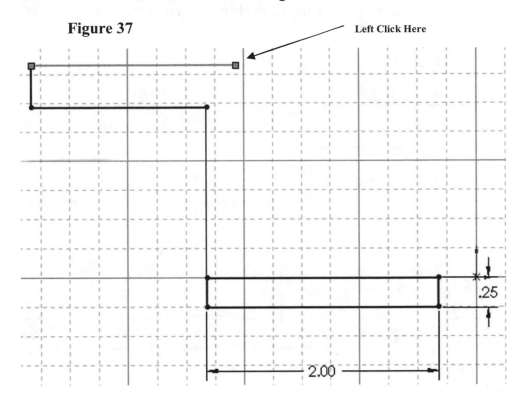

Figure 37

Left Click Here

45. With the line still attached to the cursor, move the cursor down towards the bottom of the screen and left click once as shown in Figure 38. Notice the small green square appearing at the intersection of the two lines. After the left mouse button is clicked the green square will change to a green dot. This indicates that the lines are all connected.

Figure 38

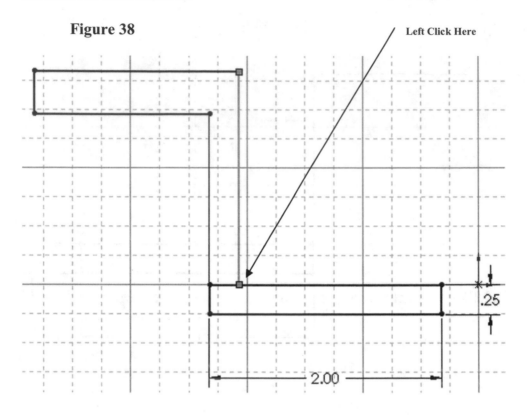

Left Click Here

.25

2.00

46. With the line still attached, right click the mouse. A pop up menu will appear. Left click on **Select** as shown in Figure 39. This verifies that no commands are active.

Figure 39

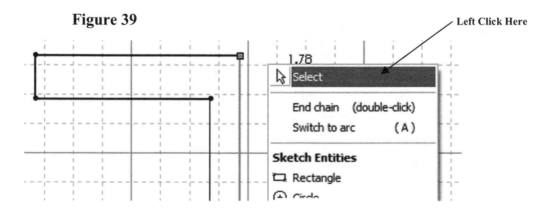

Left Click Here

47. Move the cursor to the upper far right portion of the screen and left click on the double arrows as shown in Figure 40.

Figure 40

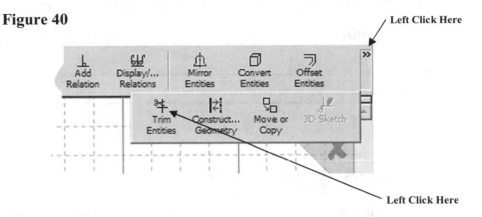

Left Click Here

Left Click Here

48. More icons will appear. Left click on **Trim Entities** as shown in Figure 40.

49. Move the cursor just above the line that will be trimmed and left click as shown in Figure 41.

Figure 41

Left Click Here

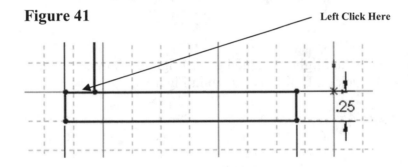

.25

50. The line will disappear as shown in Figure 42.

Figure 42

Line is Trimmed

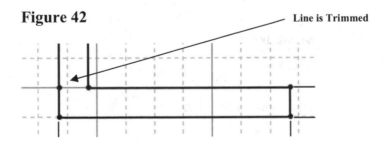

51. Move the cursor over the line in the lower left corner of the drawing as shown in Figure 43. The line will turn red. This particular line must be deleted so that the line above can be extended the full length.

Figure 43

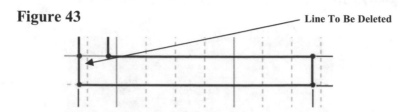

Line To Be Deleted

52. Left click once on the line that will be deleted. The line will turn green with small green squares at each end as shown in Figure 44.

Figure 44

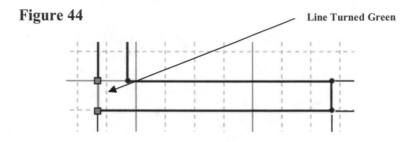

Line Turned Green

53. After the line is highlighted (turned green) right click once. A pop up menu will appear. Left click on **Delete** as shown in Figure 45.

Figure 45

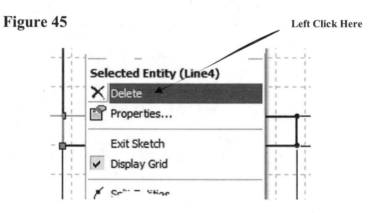

Left Click Here

Selected Entity (Line4)
X Delete
Properties...
Exit Sketch
✓ Display Grid

54. The line will be deleted as shown in Figure 46.

Figure 46

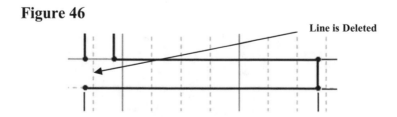

Line is Deleted

55.   Move the cursor to the upper far right portion of the screen and left click on the double arrows as shown in Figure 47.

Figure 47

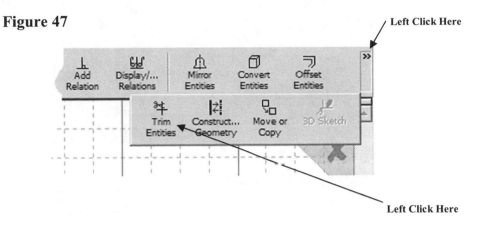

Left Click Here

Left Click Here

56.   More icons will appear. Left click on **Trim Entities** as shown in Figure 47.

57.   Move the cursor over the line to be extended. It will turn red as shown in Figure 48.

Figure 48

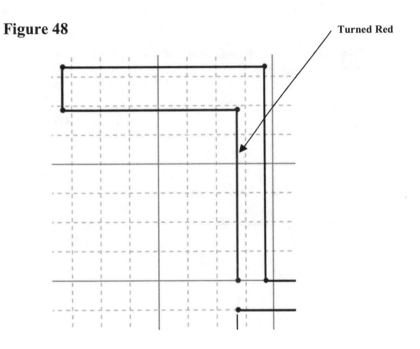

Turned Red

58. After the line has turned red, right click once. A pop up menu will appear. Left click on **Extend Entities** as shown in Figure 49.

Figure 49

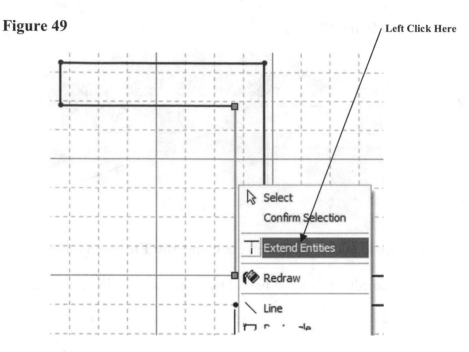

59. Move the cursor over the line to be extended. The line will extend downward forming one continuous line. Left click once as shown in Figure 50.

Figure 50

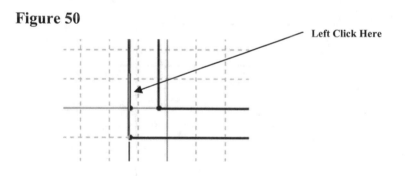

60. Your screen should look similar to Figure 51.

Figure 51

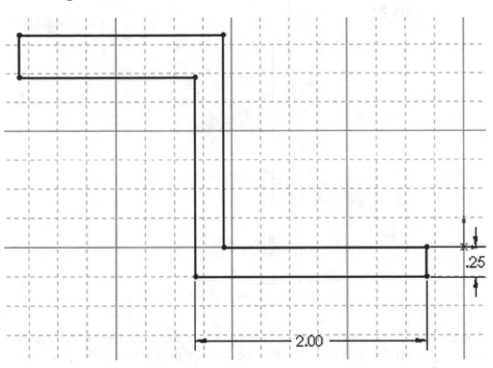

61. Move the cursor to the upper left portion of the screen and left click on **Smart Dimension** as shown in Figure 52.

Figure 52

Left Click Here

62. After selecting **Smart Dimension** move the cursor over the left side vertical line. The line will turn red as shown in Figure 53. Left click on the line.

Figure 53

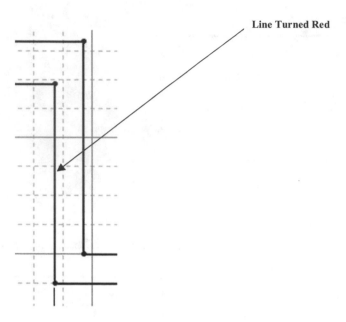

Line Turned Red

63. The dimension is attached to the cursor. Move the cursor to where the dimension will be placed and left click once. The Modify dialog box will appear as shown in Figure 54.

Figure 54

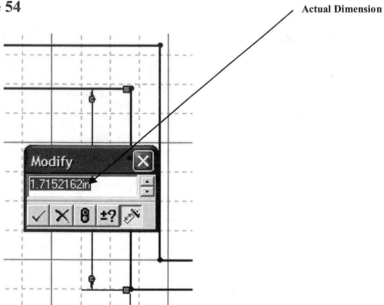

Actual Dimension

Modify

1.7152162in

64. To edit the dimension, type **1.75** in the Modify dialog box (while the current dimension is highlighted) and left click on the green checkmark as shown in Figure 55.

Figure 55

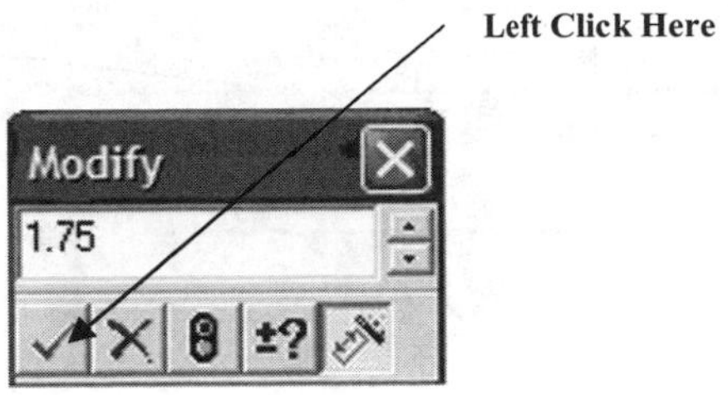

Left Click Here

65. The dimension is now 1.75 as shown in Figure 56.

Figure 56

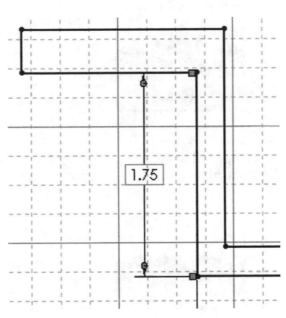

66.   Move the cursor over the left side vertical line until it turns red as shown in Figure 57. Left click on the line.

Figure 57

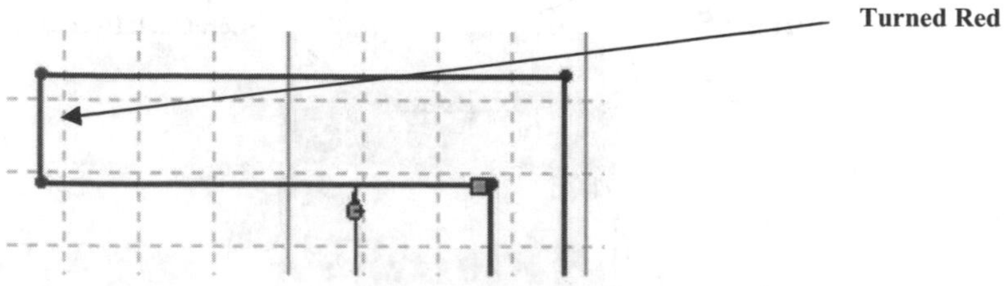

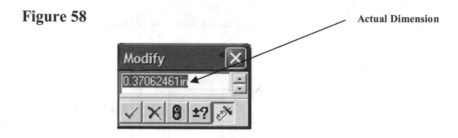

Turned Red

67.   The dimension is attached to the cursor. Move the cursor back and forth to verify it is attached. Move the cursor to where the dimension will be placed and left click once. The Modify dialog box will appear as shown in Figure 58.

Figure 58

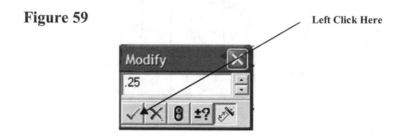

Actual Dimension

68.   To edit the dimension, type **.25** in the Modify dialog box (while the current dimension is highlighted) and left click on the green checkmark as shown in Figure 59.

Figure 59

Left Click Here

69.   The dimension is now .25 as shown in Figure 60.

Figure 60

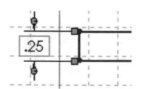

70.   Move the cursor over the top horizontal line until it turns red as shown in Figure 61. Left click on the line.

Figure 61

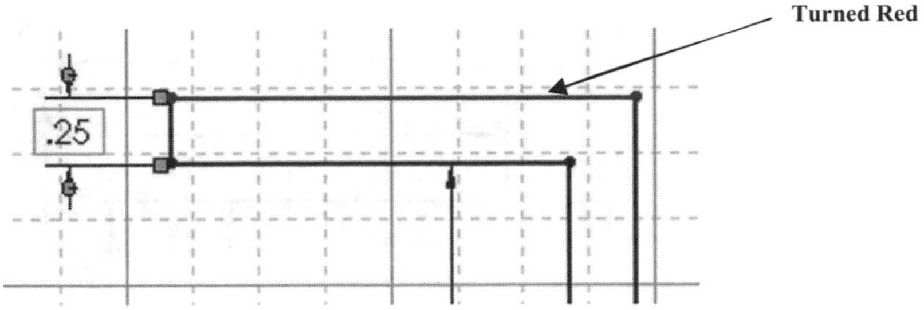

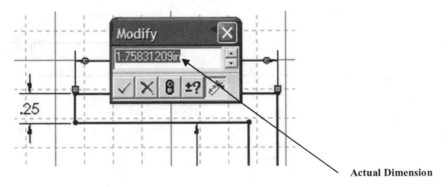

Turned Red

71.   The dimension is attached to the cursor. Move the cursor up and down to verify it is attached. Move the cursor to where the dimension will be placed and left click once. The Modify dialog box will appear as shown in Figure 62.

Figure 62

Actual Dimension

72.   To edit the dimension, type **1.75** in the Modify dialog box (while the current dimension is highlighted) and left click on the green checkmark as shown in Figure 63.

Figure 63

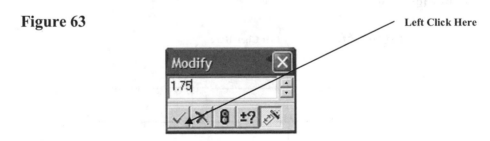

Left Click Here

73. The dimension is now 1.75 as shown in Figure 64.

Figure 64

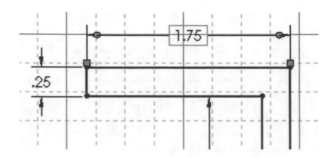

74. Move the cursor over to the left side vertical line until it turns red and left click as shown in Figure 65.

Figure 65

Left Click Here

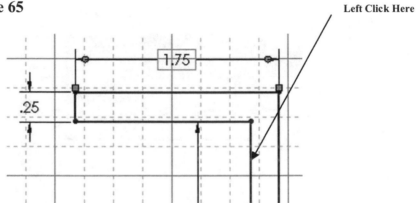

75. Move the cursor to the other vertical line until it turns red and left click as shown in Figure 66.

Figure 66

Left Click Here

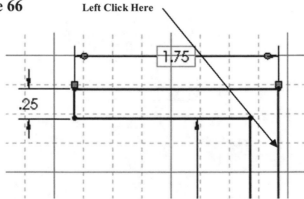

76. The dimension is attached to the cursor. Move the cursor back and forth to verify it is attached. Move the cursor to where the dimension will be placed and left click once. While the dimension is highlighted, left click the mouse once. The Modify dialog box will appear as shown in Figure 67.

Figure 67

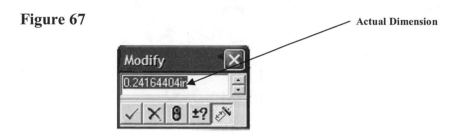

Actual Dimension

77. To edit the dimension, type **.25** in the Modify dialog box (while the current dimension is highlighted) and left click on the green checkmark as shown in Figure 68.

Figure 68

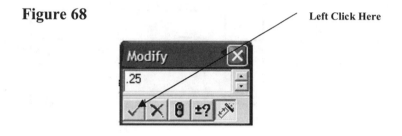

Left Click Here

39

78. Your screen should look similar to Figure 69.

Figure 69

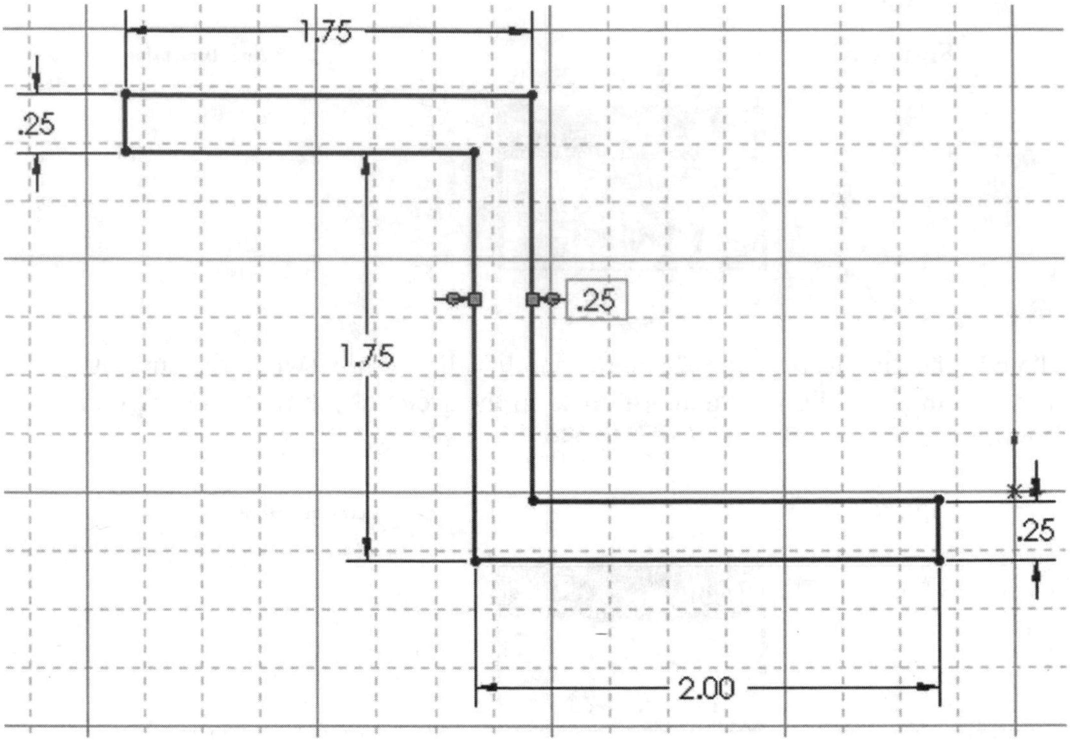

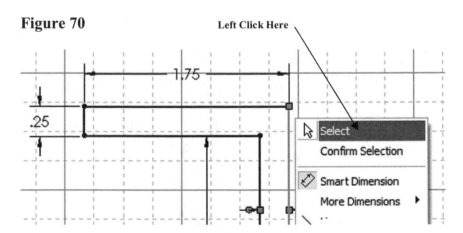

79. Right click near the drawing. A pop up menu will appear. Left click on **Select** as shown in Figure 70. This will ensure that no commands are still active.

Figure 70

80. Right click anywhere on the drawing. A pop up menu will appear. Left click on **Exit Sketch** as shown in Figure 71.

Figure 71

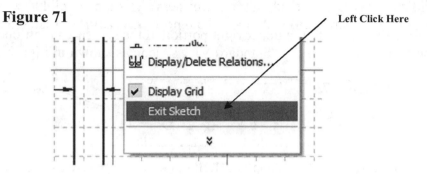

Left Click Here

81. Your screen should look similar to Figure 72.

Figure 72

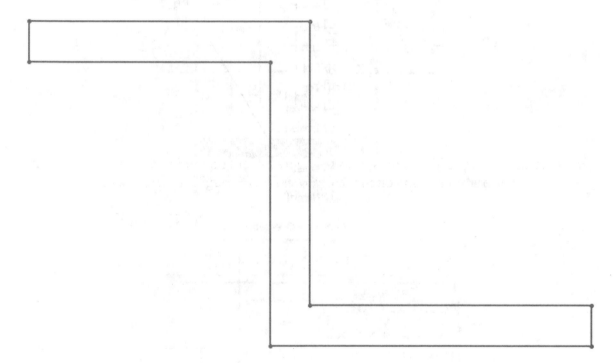

82. To utilize the Features commands, a sketch must be present and have no opens (non-connected lines). If there are any opens in the sketch an error message will appear.

83. Move the cursor to the upper right portion of the screen and left click on the drop down arrow next to the "Standard Views" icon as shown in Figure 73.

Figure 73

Left Click Here

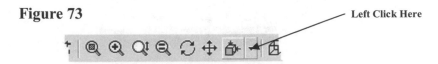

84. A drop down menu will appear. Left click on **Trimetric** as shown in Figure 74.

Figure 74

Left Click Here

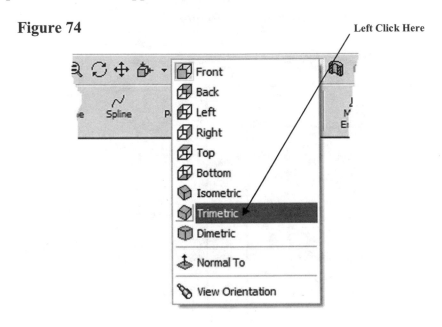

85. The view will become trimetric as shown in Figure 75.

Figure 75

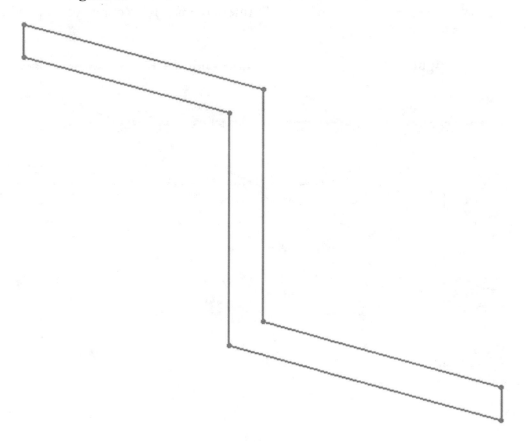

86. Move the cursor to the middle left portion of the screen and left click on **Extruded Boss/Base** as shown in Figure 76.

Figure 76　　　　　　　　　　　　　　　　Left Click Here

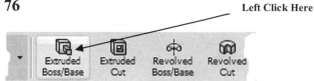

87.   The Extrude dialog box will appear. SolidWorks also provides a preview of the extrusion. If SolidWorks gave you an error message, there are opens (non-connected lines) somewhere on the sketch. Check each intersection for opens by using the Extend Entities and Trim Entities commands. Your screen should look similar to Figure 77.

Figure 77

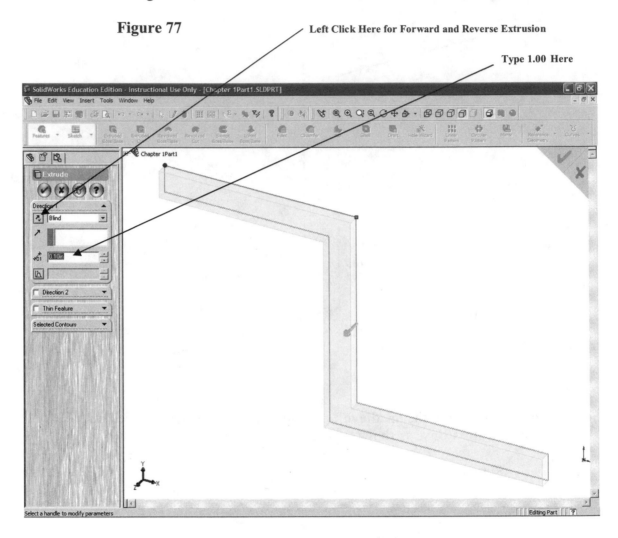

88. Left click on the icon under Direction, located next to the text "Blind". The extrusion will change directions. Ensure that the extrusion is facing forward. While the text located next to "D1" is still highlighted, enter **1.000.** To preview the extruded part, left click on the "Glasses" icon located at the top of the Extrude dialog box as shown in Figure 78. The part will become purple.

Figure 78

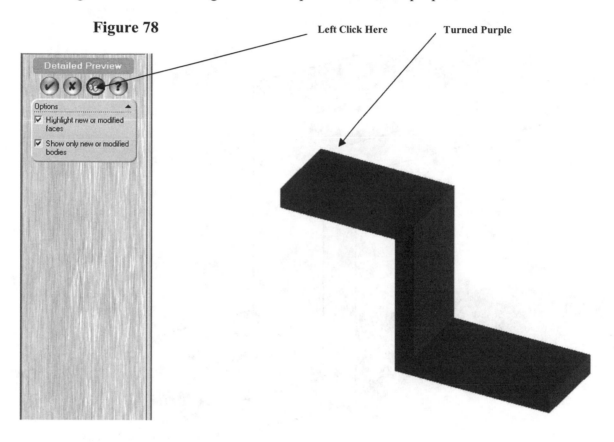

89. To end preview, left click on the "Glasses" icon again. This will return the part to the previous view.

90. Left click on the green checkmark as shown in Figure 79.

Figure 79

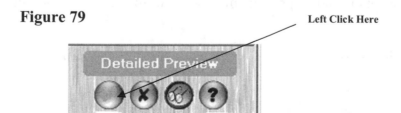

91.   Left clicking on the green checkmark will create a solid from the sketch. Your screen should look similar to Figure 80.

Figure 80

92.   Move the cursor to the upper middle portion of the screen and left click on **Fillet** as shown in Figure 81.

Figure 81   Left Click Here

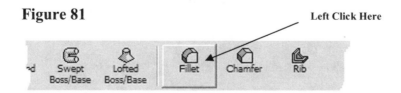

93. The menu at the left will change. Enter **.5** for the radius and select **Full Preview** as shown in Figure 82.

Figure 82

Enter .5 Here

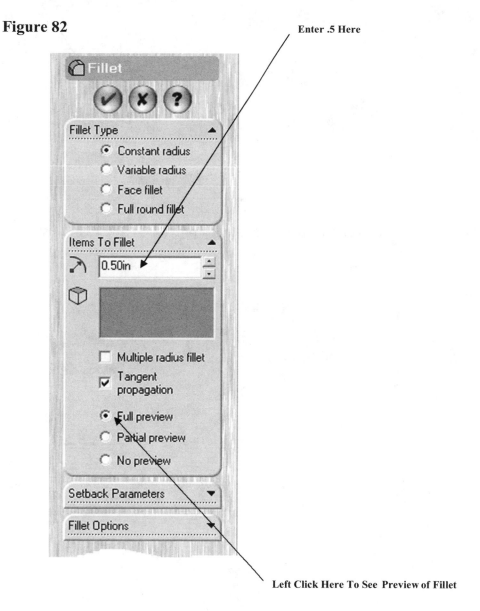

Left Click Here To See Preview of Fillet

94. Move the cursor over the lower front edge causing it to turn red and left click once as shown in Figure 83.

Figure 83

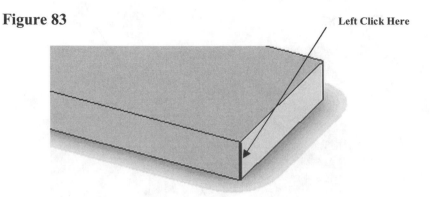

Left Click Here

95. Notice the yellow lines illustrating a preview of the fillet as shown in Figure 84.

Figure 84

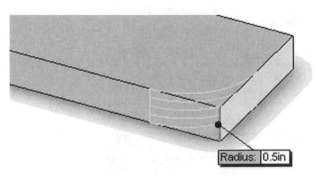

96. Move the cursor over the upper front edge causing it to turn red and left click once as shown in Figure 85.

Figure 85

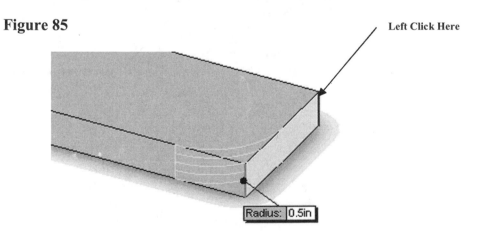

Left Click Here

97. Notice the yellow lines illustrating a preview of the fillet as shown in Figure 86.

Figure 86

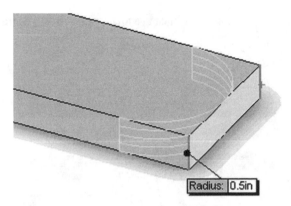

98. Left click on the two upper remaining edges. You may have to use the rotate command to select the upper edge that is not visible.

99. Move the cursor to the upper middle portion of the screen and left click on the "Rotate" icon as shown in Figure 87.

Figure 87   Left Click Here

100. Left click on the left portion of the part (holding the left mouse button down). Drag the cursor to the right allowing access to the upper right corner of the part as shown in Figure 88.

Figure 88 Left Click Here, Hold Left Mouse Button Down, and Drag to the Right

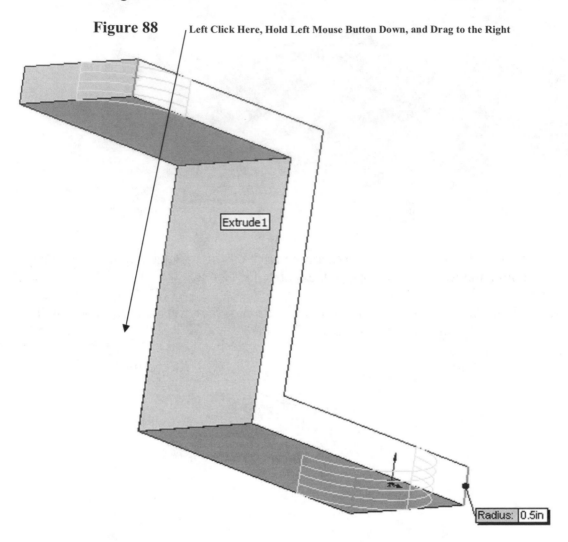

Extrude1

Radius: 0.5in

101. Move the cursor to the upper middle portion of the screen and left click on the "Rotate" icon again as shown in Figure 89.

Figure 89 Left Click Here

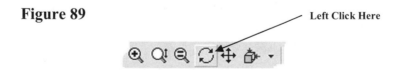

102. Move the cursor to the upper right corner of the part causing it to turn red. Left click once as shown in Figure 90.

Figure 90

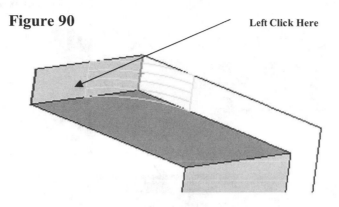

Left Click Here

103. Your screen should look similar to Figure 91.

Figure 91

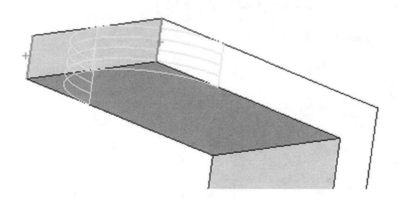

104. Move the cursor to the upper right portion of the screen and left click on the drop down arrow next to the "Standard Views" icon as shown in Figure 92.

Figure 92

Left Click Here

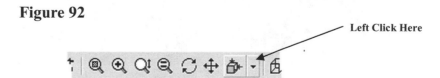

105. A drop down menu will appear. Left click on **Trimetric** as shown in Figure 93.

Figure 93

Left Click Here

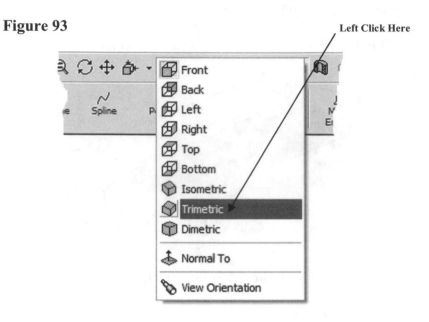

106. Your screen should look similar to Figure 94.

Figure 94

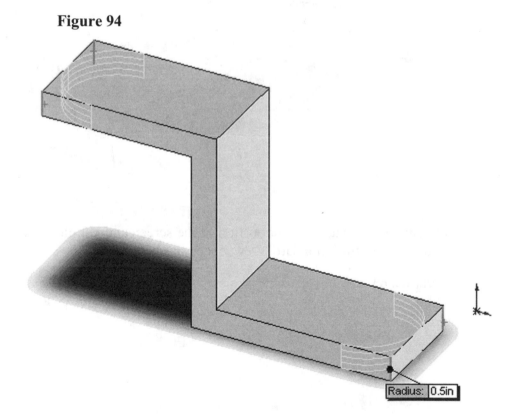

107.   Left click on the green checkmark as shown in Figure 95.

Figure 95

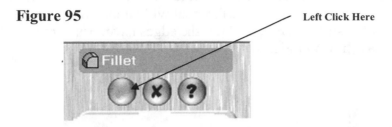

Left Click Here

108.   Your screen should look similar to Figure 96.

Figure 96

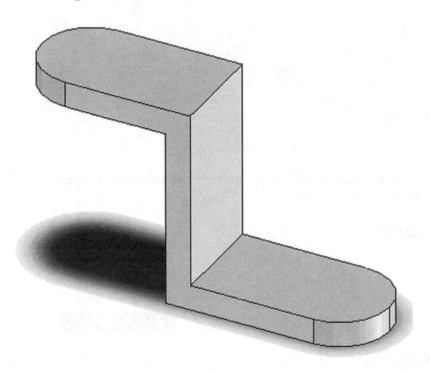

109.   The next task will include cutting a hole in each of the ends. To do this, a sketch
       will need to be constructed on each surface. To begin a new sketch on any
       surface, move the cursor to the surface that will include the new sketch. Notice
       the edges of the surface are red. After the edges have turned red, right click on
       the surface as shown in Figure 97.

Figure 97

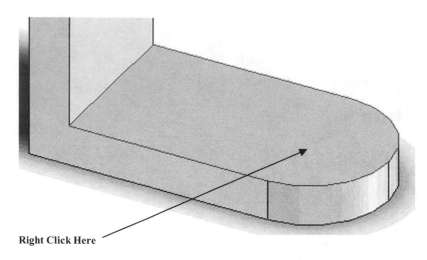

Right Click Here

110.   The surface will change color. A pop up menu will appear. Left click on **Insert
       Sketch** as shown in Figure 98.

Figure 98

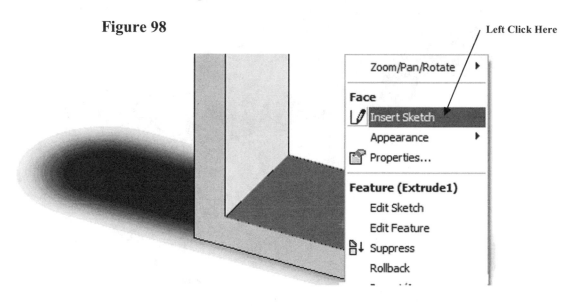

111.   SolidWorks will create a "sketch" on that particular surface. Notice the toolbar at
       the top of the screen has changed back to the sketch commands.

112. Your screen should look similar to Figure 99.

Figure 99

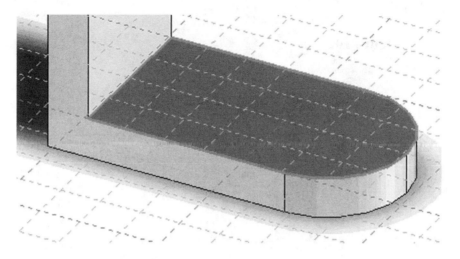

113. Move the cursor to the upper left portion of the screen and left click on **Circle** as shown in Figure 100.

Figure 100   Left Click Here

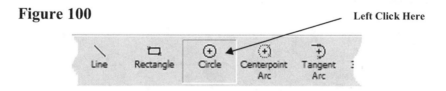

114. Move the cursor to the edge of the circle. The center point of the circle will appear as shown in Figure 101.

Figure 101   Move Cursor Here

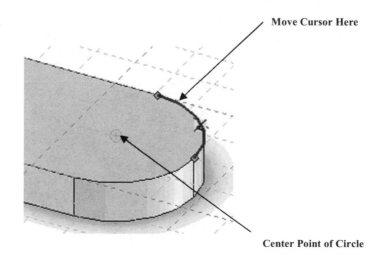

Center Point of Circle

115. After the center marker appcars, left click once. This will be the center of a circle, which will later become a thru hole. Move the cursor out to the side to make the hole larger. Move the cursor out far enough to create a hole size similar to Figure 102.

Figure 102

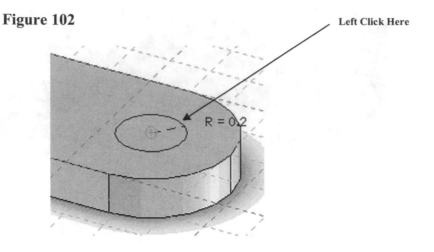

116. After the hole size looks similar to Figure 102, left click once.

117. Move the cursor to the upper left portion of the screen and left click on **Smart Dimension** as shown in Figure 103.

Figure 103

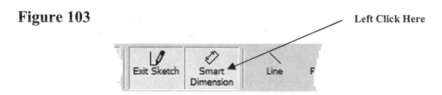

118. Left click on the edge (not the center) of the circle as shown in Figure 104. The circle will turn red.

Figure 104

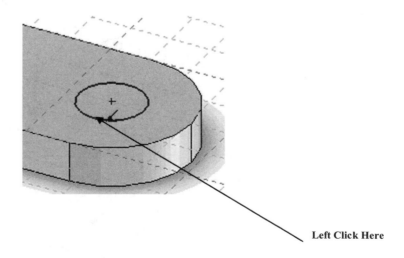

119. The dimension will appear attached to the cursor. Move the cursor up and down to verify it is attached. Move the cursor to where the dimension will be placed and left click once as shown in Figure 105.

Figure 105

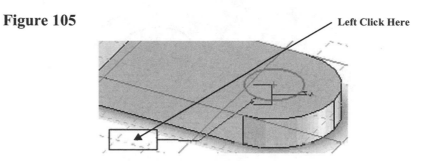

120. The Modify dialog box will appear as shown in Figure 106.

Figure 106

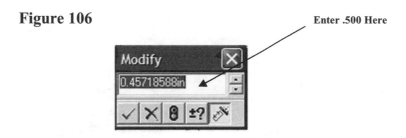

121. Type **.500** in the Modify dialog box and left click on the green checkmark as shown in Figure 107.

Figure 107

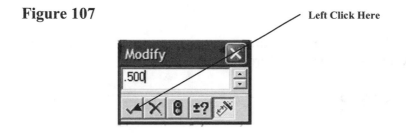

122. The diameter of the hole will become .500 inches as shown in Figure 108.

Figure 108

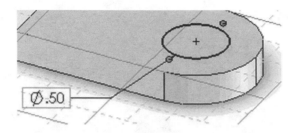

123. Right click near the drawing. A pop up menu will appear. Left click on **Select** as shown in Figure 109. This will ensure that no commands remain active.

Figure 109

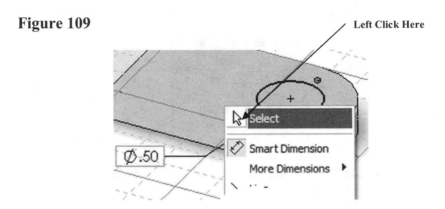

124. Right click anywhere on the drawing. A pop up menu will appear. Left click on **Exit Sketch** as shown in Figure 110.

Figure 110

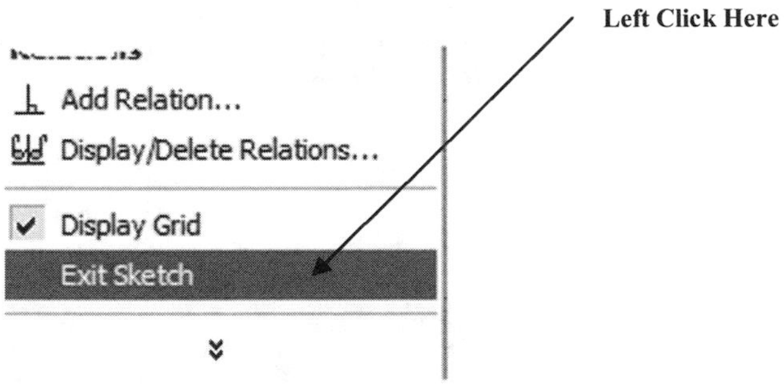

125. SolidWorks is now out of the Sketch commands and into the Features Commands. Notice that the commands at the top of the screen are now different as shown in Figure 111.

Figure 111

126. Move the cursor to the upper middle portion of the screen and left click on **Extruded Cut** as shown in Figure 112. This time SolidWorks will extrude "space" or "air" rather than material as was done to create the bracket.

Figure 112   Left Click Here

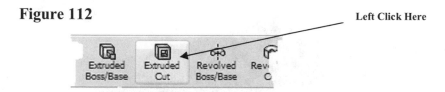

127. The menu at the left will change. Enter **.500** as shown in Figure 113.

Figure 113   Enter .500 Here

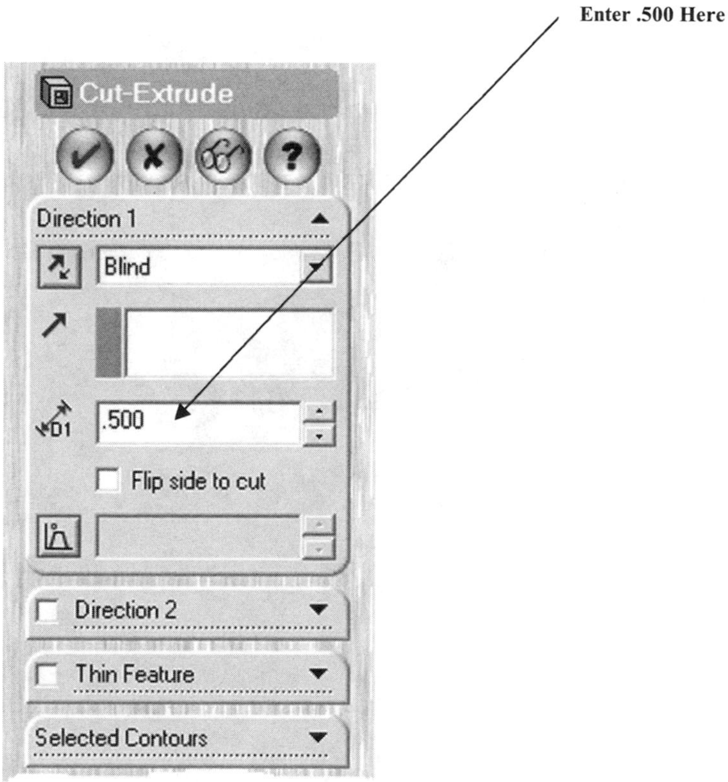

128.   Move the cursor to the upper left portion of the screen and left click on the "Glasses" icon as shown in Figure 114.

Figure 114

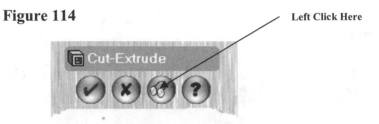

Left Click Here

129.   A preview of the part will be displayed as shown in Figure 115.

Figure 115

130.   To return to the main menu, left click once on the "Glasses" icon.

131.   Left click on the green checkmark as shown in Figure 116.

Figure 116

Left Click Here

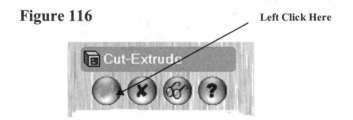

132. There should be a thru hole in the part similar to Figure 117.

Figure 117

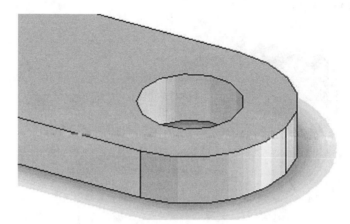

133. Another method of creating a hole is to use the Hole Wizard command.

134. To use the Hole Wizard command a "Point" must be constructed in the Sketch command. Move the cursor to the top portion of the part as shown in Figure 118. The outer edges of the part will turn red.

Figure 118

Move Cursor Here

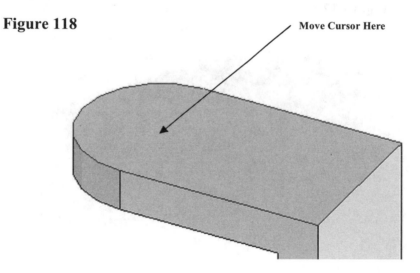

135. Right click on the surface. The surface will change color. A pop up menu will also appear as shown in Figure 119.

Figure 119

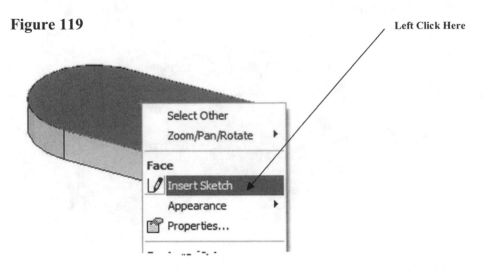

Left Click Here

136. Left click on **Insert Sketch**. SolidWorks will return to the Sketch commands as shown in Figure 120.

Figure 120

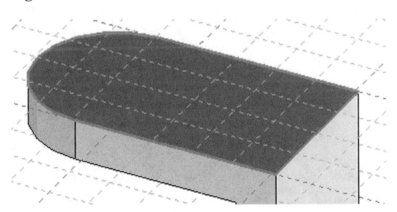

137. Move the cursor to the upper right portion of the screen and left click on **Point** as shown in Figure 121.

Figure 121

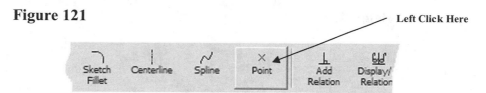

Left Click Here

138. Move the cursor to edge of the part. A portion of the edge will turn red. The center point of the fillet radius will also appear. Left click once on the center point as shown in Figure 122.

Figure 122

Left Click Here

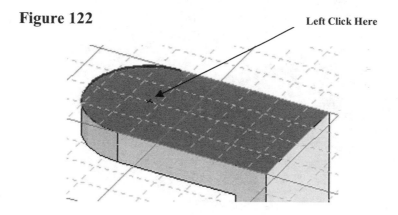

139. A small center marker will appear on the center of the fillet radius as shown in Figure 123.

Figure 123

Center Point of Radius

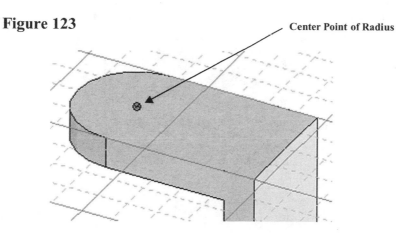

140.   Move the cursor to the upper left portion of the screen and left click on the green checkmark as shown in Figure 124.

Figure 124

Left Click Here

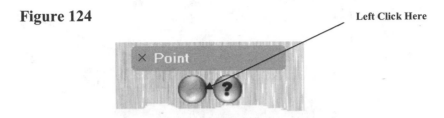

141.   Right click anywhere on the drawing. A pop up menu will appear. Left click on **Select** as shown in Figure 125.

Figure 125

Left Click Here

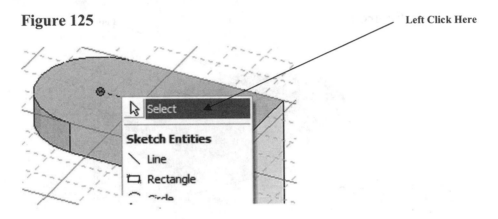

142.   Right click anywhere on the drawing. A pop up menu will appear. Left click on **Exit Sketch** as shown in Figure 126.

Figure 126

Left Click Here

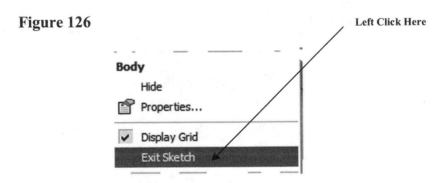

143. SolidWorks is now out of the Sketch commands and into the Features commands. Notice that the commands at the top of the screen are now different. Your screen should look similar to Figure 127.

Figure 127

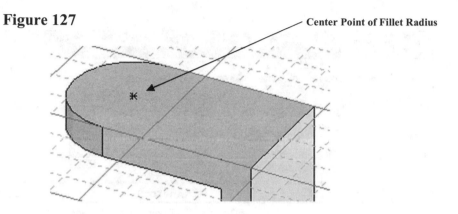

Center Point of Fillet Radius

144. Right click anywhere around the part. A pop up menu will appear. Left click on **Exit Sketch** as shown in Figure 128.

Figure 128

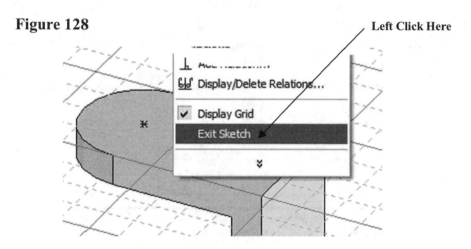

Left Click Here

145. Move the cursor to the upper left portion of the screen and left click on **Hole Wizard** as shown in Figure 129.

Figure 129

Left Click Here

146. The Hole Definition dialog box will appear. Enter **.25** for Hole Fit and Diameter, **.500** for C'Bore Diameter and **.125** for the C'Bore Depth as shown in Figure 130.

Figure 130

Enter .25 Here

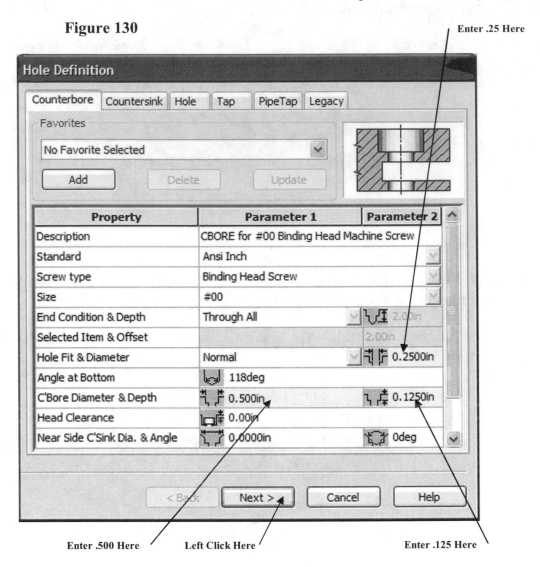

Enter .500 Here   Left Click Here   Enter .125 Here

147. Left Click on **Next** as shown in Figure 130.

148. The Hole Placement dialog box will appear. Move the cursor to the center point of the fillet radius and left click once as shown in Figure 131.

Figure 131

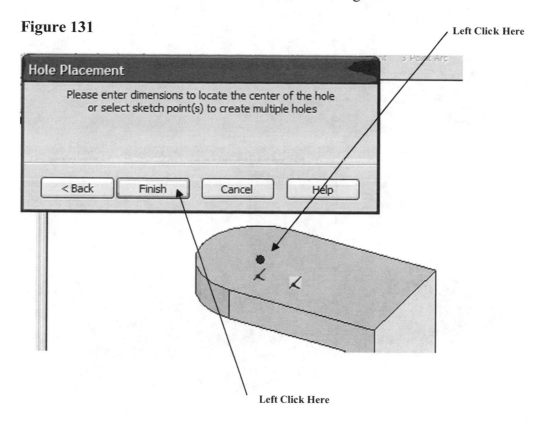

149. Left click on **Finish**.

150. Your screen should look similar to Figure 132.

Figure 132

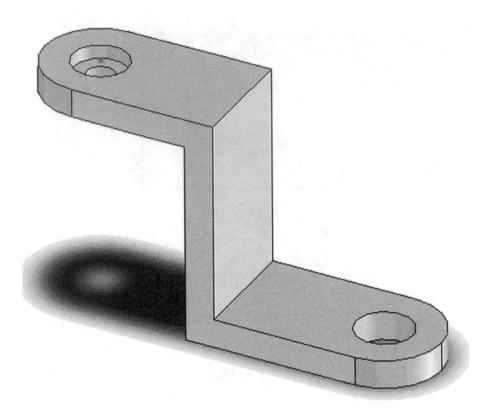

151. To ensure that the hole is correct move the cursor to the top portion of the screen and left click on the "Rotate" icon as shown in Figure 133.

Figure 133   Left Click Here

152. Left click anywhere around the part and hold the left mouse button down. Drag the cursor upward. The part will rotate upward as shown in Figure 134.

Figure 134

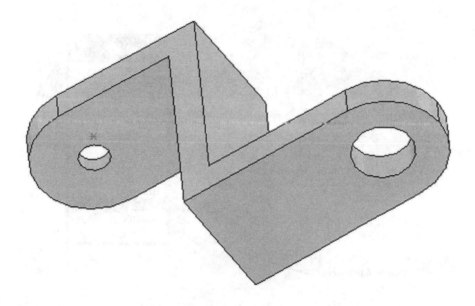

153. Holding the left mouse button down keeps the part attached to the cursor. To view the part in Trimetric, right click anywhere on the screen and left click on **View Orientation** as shown in Figure 135.

Figure 135

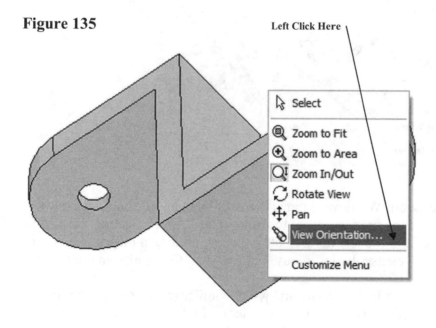

154. A drop down menu will appear. Double click on **Trimetric View** as shown in Figure 136.

Figure 136

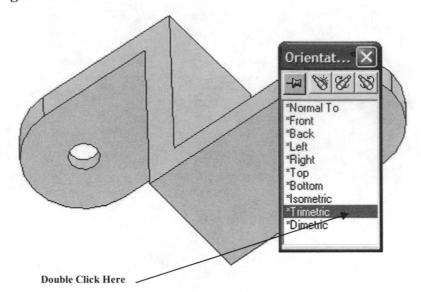

Double Click Here

155. Other commands for viewing are located at the top of the screen as shown in Figure 137.

Figure 137

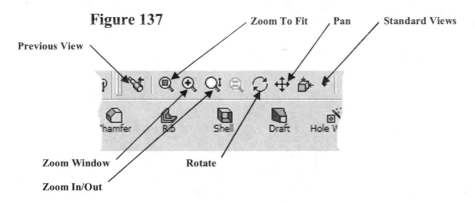

156. The Zoom Window command works by using the cursor to draw a window around the area you want to zoom in on. After selecting the "Zoom Window" icon, hold the left mouse button down and drag a diagonal box around the desired area. Release the left mouse button when the proper amount of zoom is achieved.

157. The Zoom In/Out command works similar to the Zoom Window command. Start by selecting the "Zoom In/Out" icon. Left click on the drawing and hold the left mouse button down while dragging the cursor up and down until the proper amount of zoom is achieved.

158. The Pan command works similar to the Zoom In/Out command. Start by selecting the "Pan" icon. Left click on the drawing and hold the left mouse button down while moving the cursor up and down or side to side. Release the mouse button after the desired view is achieved.

159. The Zoom to Fit command works by filling the screen so that the entire part is as large as possible while remaining visible in its entirety.

160. The same viewing options can be accessed by right clicking anywhere on the drawing. A pop will be displayed as shown in Figure 138.

Figure 138

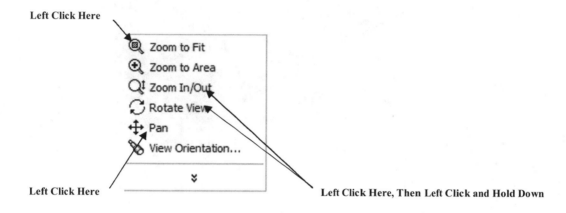

Drawing Activities

Problem 1

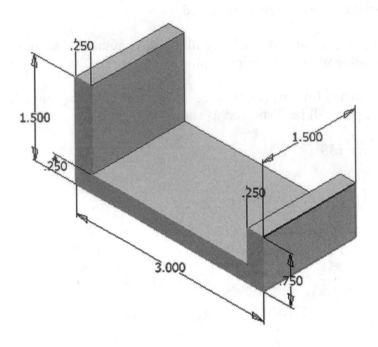

Problem 2

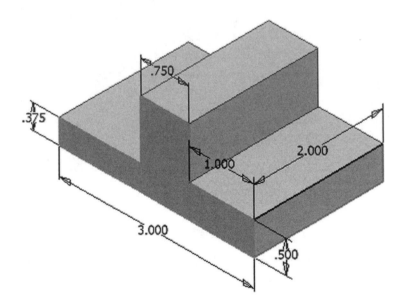

Problem 3

Extrude Center Section .25 Deep

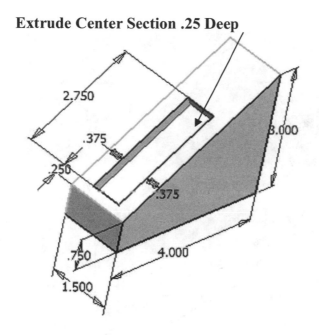

Problem 4

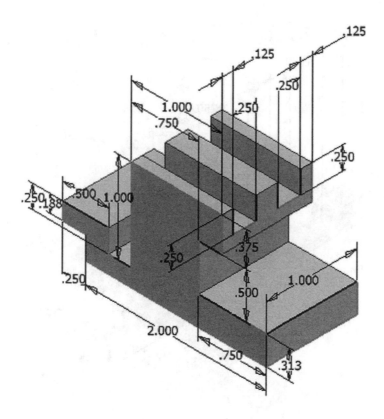

Problem 5

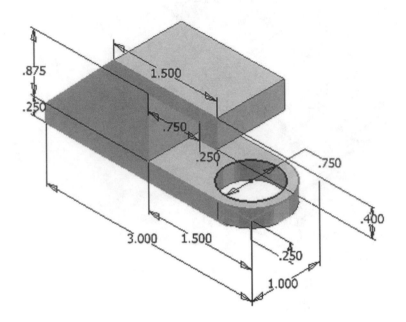

Problem 6

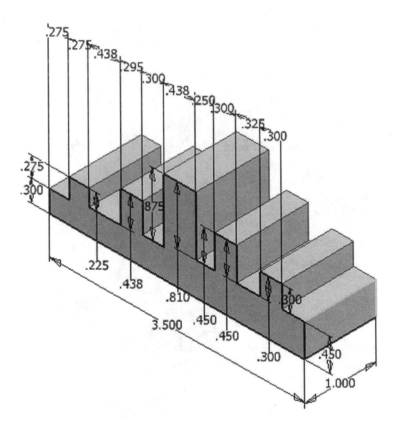

Problem 7

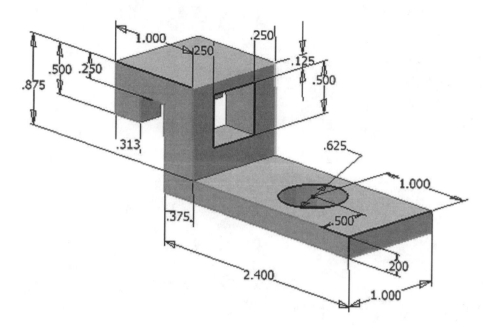

Problem 8

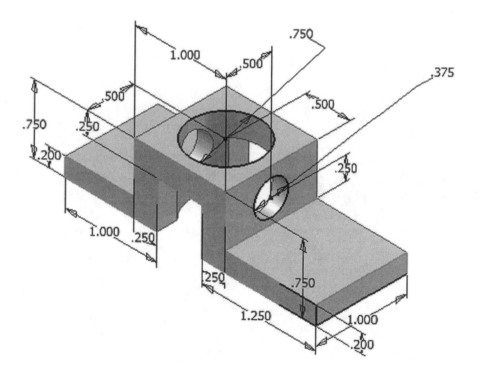

Chapter 2 Learning More Basics

Objectives:

- Create a simple sketch using the Sketch commands
- Dimension a sketch using the Smart Dimension command
- Revolve a sketch using the Revolve command
- Create a hole using the Extrude command
- Create a series of holes using the Circular Hole command

Chapter 2 includes instruction on how to design the part shown below.

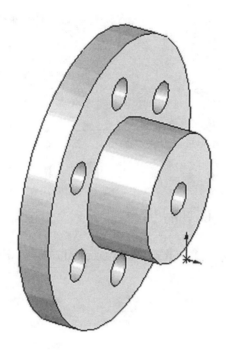

1.   Start SolidWorks by referring to "Chapter 1 Getting Started".

2.   After SolidWorks is running, begin a new sketch.

3.   Move the cursor to the upper left corner of the screen and left click on **Line** as shown in Figure 1.

Figure 1   Left Click Here

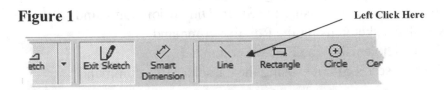

4.   Move the cursor to the lower left portion of the screen and left click once. This will be the beginning end point of a line as shown in Figure 2.

Figure 2   Beginning Endpoint of Line

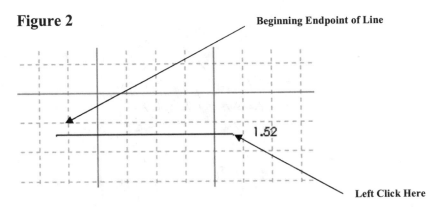

                                                                Left Click Here

5.   Move the cursor to the right and left click once as shown in Figure 2.

6.   Move the cursor up and left click once as shown in Figure 3.

Figure 3   Left Click Here

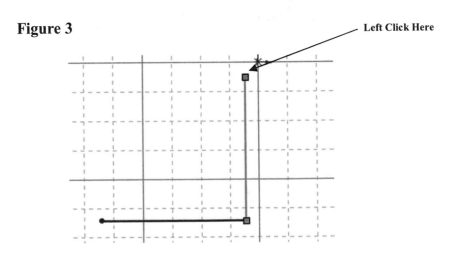

7.   Move the cursor to the right and left click once as shown in Figure 4.

Figure 4

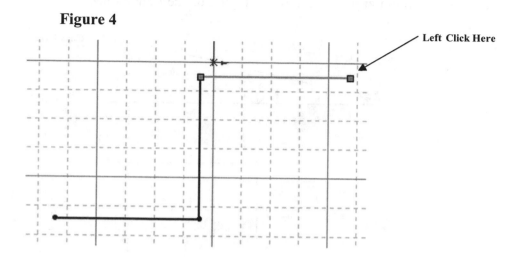

8.   Move the cursor up and left click once as shown in Figure 5.

Figure 5

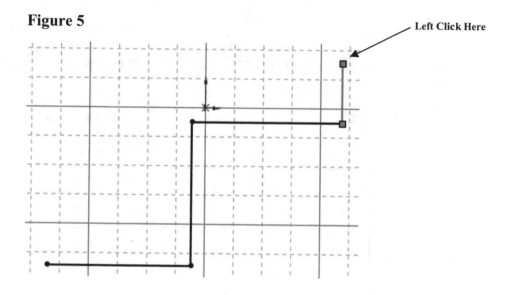

9. Move the cursor to the left. Ensure that the dots between the first end point and the last end point appear as shown in Figure 6. Left click once.

Figure 6

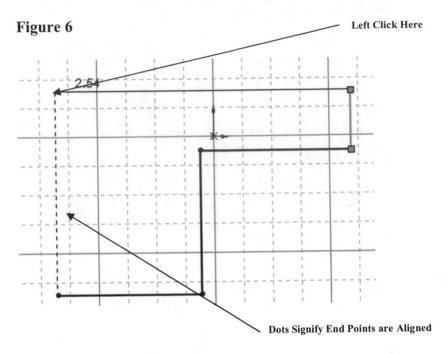

Left Click Here

2.54

Dots Signify End Points are Aligned

10. Move the cursor back to the original starting end point and left click once as shown in Figure 7.

Figure 7

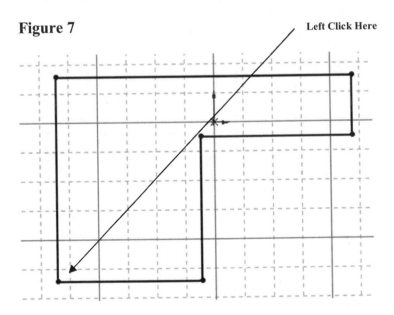

Left Click Here

11. Move the cursor to the middle left portion of the screen and left click on **Smart Dimension** as shown in Figure 8.

Figure 8

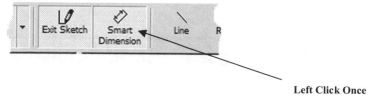

Left Click Once

12. After selecting **Smart Dimension** move the cursor over the bottom horizontal line until it turns red as shown in Figure 9. Select the line by left clicking anywhere on the line **or** on each of the end points. To use the end points of the line, move the cursor over one of the end points. A small red dot will appear. Left click once and move the cursor to the other end point. After the red dot appears, left click once. The dimension will be attached to the cursor.

Figure 9

Turned Red

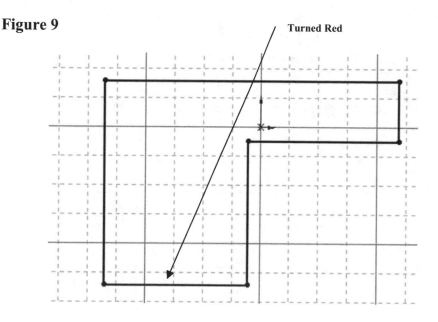

13. Move the cursor to where the dimension will be placed and left click once as shown in Figure 10.

Figure 10

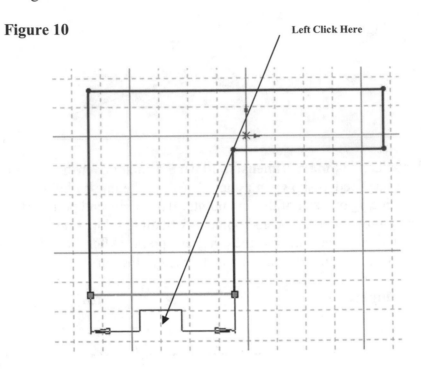

14. The Modify dialog box will appear as shown in Figure 11.

Figure 11

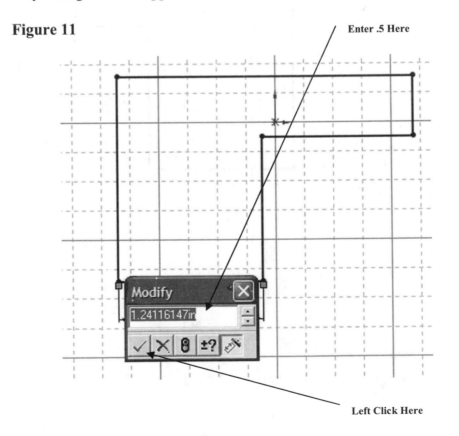

15.   To edit the dimension, type **.5** in the Modify dialog box (while the current dimension is highlighted) and either press **Enter** on the keyboard or left click on the green checkmark as shown in Figure 11.

16.   The dimension of the line will become .50 inches as shown in Figure 12.

Figure 12

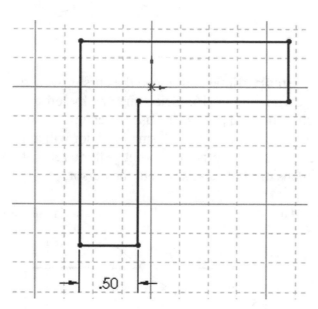

17.   To view the entire drawing move the cursor to the middle portion of the screen and left click once on the "Zoom To Fit" icon as shown in Figure 13.

Figure 13   Left Click Here

18.   The drawing will "fill up" the entire screen. If the drawing is still too large, left click on the "Zoom In/Out" icon as shown in Figure 14. After selecting the icon, hold the left mouse button down and move the cursor up and down to achieve the desired view of the sketch.

Figure 14   Left Click Here

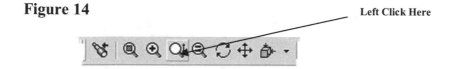

19.   Move the cursor to the middle left portion of the screen and left click on **Smart Dimension** as shown in Figure 15.

Figure 15

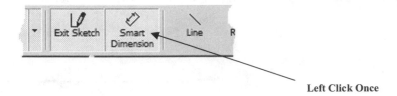

Left Click Once

20.   After selecting **Smart Dimension** move the cursor over the vertical line until it turns red as shown in Figure 16. Select the line by left clicking anywhere on the line **or** on each of the end points. To use the end points of the line, move the cursor over one of the end points. A small red dot will appear. Left click once and move the cursor to the other end point. After a red dot appears, left click once. The dimension will be attached to the cursor.

Figure 16   Turned Red

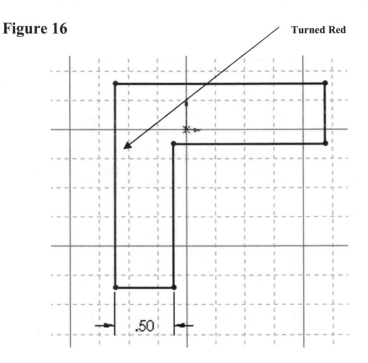

21.   Move the cursor to where the dimension will be placed and left click once as shown in Figure 17.

Figure 17

Left Click Here

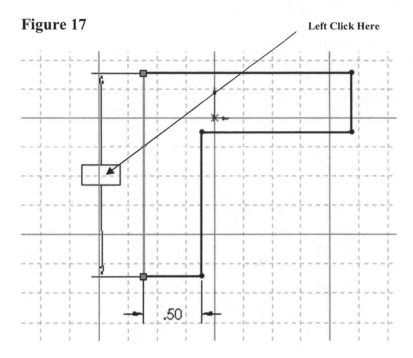

22.   The Modify dialog box will appear as shown in Figure 18.

Figure 18

Enter 2.0 Here   Left Click Here

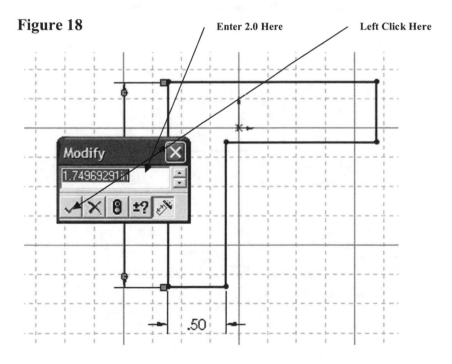

23. To edit the dimension, type **2.0** in the Modify dialog box (while the current dimension is highlighted) and either left click on the green checkmark (as shown in Figure 18) or press **Enter** on the keyboard.

24. The dimension of the line will become 2.0 inches as shown in Figure 19. Use the Zoom icons to zoom out if necessary.

Figure 19

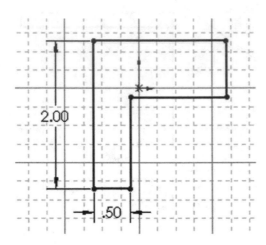

25. Select the next line by left clicking anywhere on the line **or** on each of the end points. The dimension will be attached to the cursor.

Figure 20   Line Turned Red

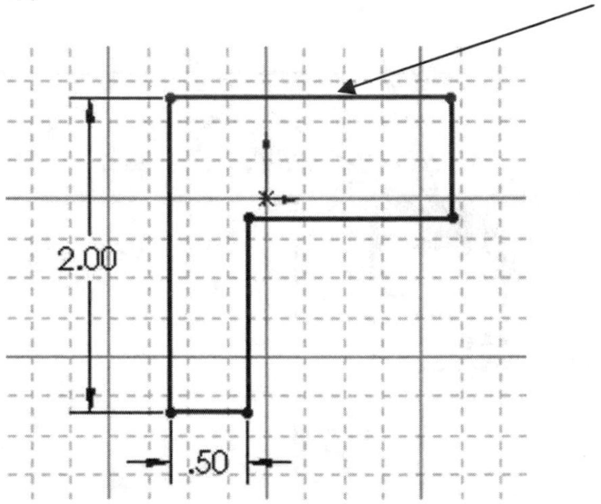

26. Move the cursor to where the dimension will be placed and left click once as shown in Figure 21.

Figure 21

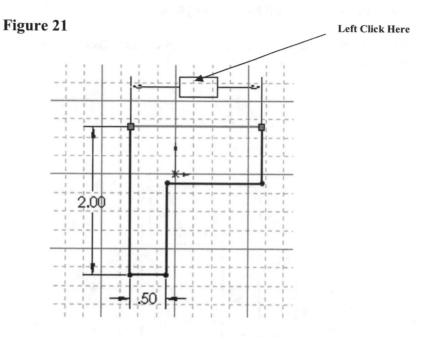

27. The Modify dialog box will appear as shown in Figure 22.

Figure 22

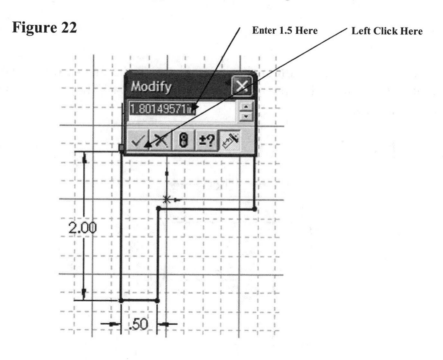

28.   To edit the dimension, type **1.5** in the Modify dialog box (while the current dimension is highlighted) and either left click on the green checkmark (as shown in Figure 22) or press **Enter** on the keyboard.

29.   The dimension of the line will become **1.5** inches as shown in Figure 23. Use the Zoom icons to zoom out if necessary.

Figure 23

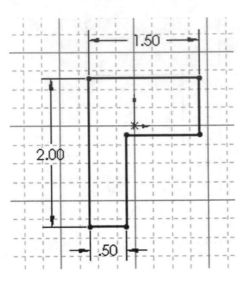

30.   Select the next line by left clicking anywhere on the line **or** on each of the end points. The line will turn red as shown in Figure 24. The dimension will be attached to the cursor.

Figure 24   Turned Red

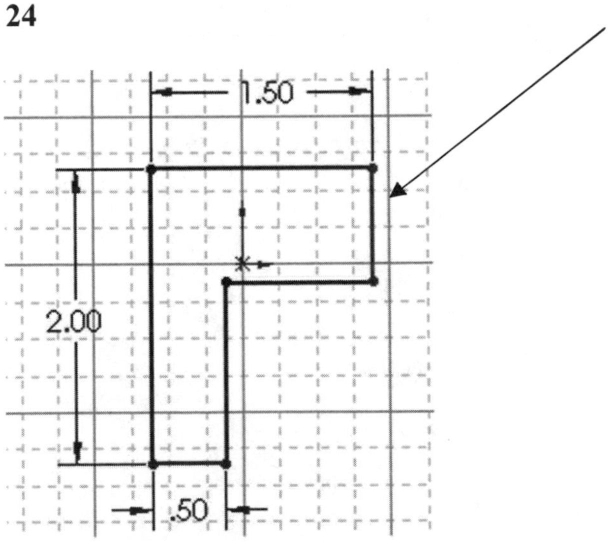

31. Move the cursor to where the dimension will be placed and left click once as shown in Figure 25.

Figure 25

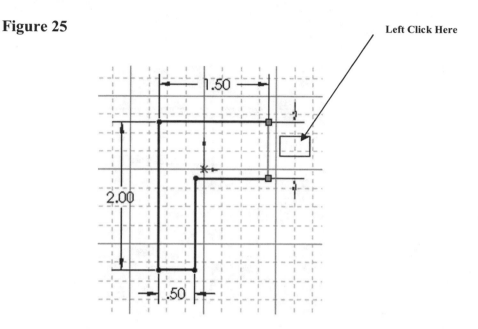

32. The Modify dialog box will appear as shown in Figure 26.

Figure 26

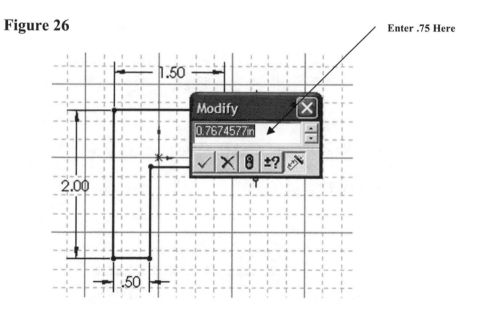

33. To edit the dimension, type **.75** in the Modify dialog box (while the current dimension is highlighted) and either click on the green checkmark or press **Enter** on the keyboard.

34. The dimension of the line will become **.75** inches as shown in Figure 27. Use the Zoom icons to zoom out if necessary.

Figure 27

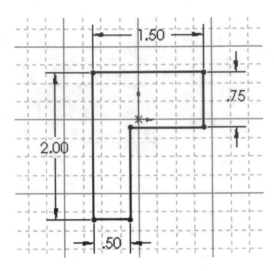

35. Move the cursor to the upper left corner of the screen and left click on **Line** as shown in Figure 28.

Figure 28   **Left Click Here**

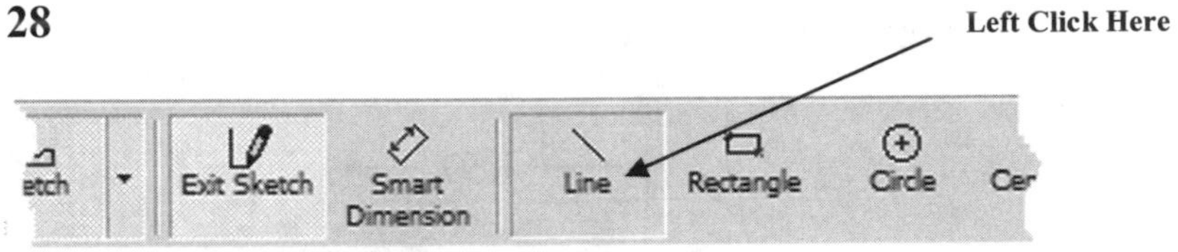

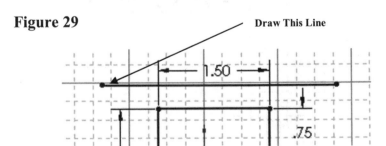

36. Draw a line parallel to the top horizontal line as shown in Figure 29.

Figure 29   **Draw This Line**

37. Move the cursor to where the dimension will be placed and left click once. The Modify dialog box will appear. To edit the dimension, type **.25** in the Modify dialog box (while the current dimension is highlighted) and either click on the green checkmark or press **Enter** on the keyboard. Your screen should look similar to Figure 30.

Figure 30

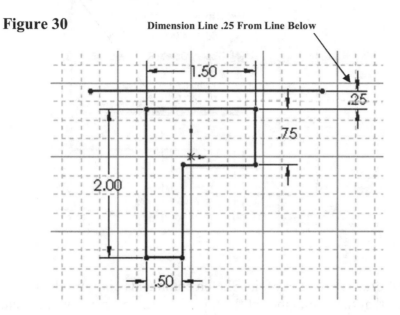

Dimension Line .25 From Line Below

38. Right click anywhere on the drawing. A pop up menu will appear. Left click on **Select** as shown in Figure 31. This verifies that no commands are active.

Figure 31

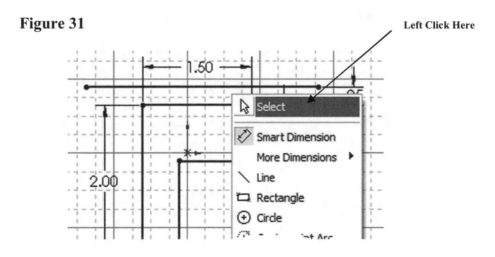

Left Click Here

39. After the sketch is complete it is time to revolve the sketch into a solid.

40. Right click anywhere on the sketch. A pop up menu will appear. Left click on **Exit Sketch** as shown in Figure 32.

Figure 32

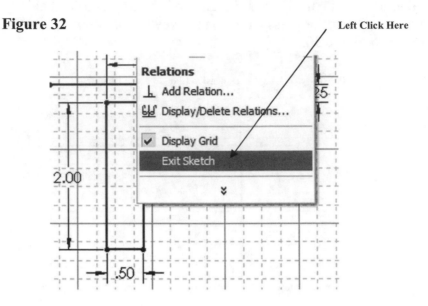

41. SolidWorks is now out of the Sketch commands and into the Features commands. Notice that the commands at the top of the screen are now different. To work in the Features commands a sketch must be present and have no opens (non-connected lines). If there are any opens in the sketch an error message will appear. Your screen should look similar to Figure 33.

Figure 33

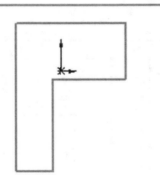

42. Move the cursor to the upper right portion of the screen and left click on the drop down arrow next to the "Standard Views" icon. A drop down menu will appear as shown in Figure 34.

Figure 34

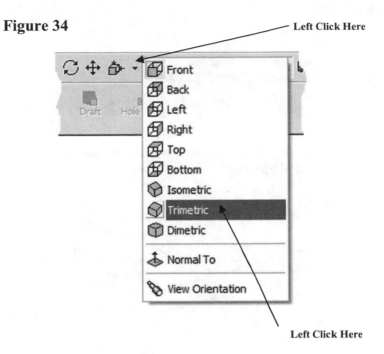

43. Left click on **Trimetric** as shown in Figure 34.

44. The view will become trimetric as shown in Figure 35.

Figure 35

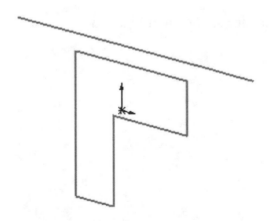

45. Move the cursor to the upper left portion of the screen and left click on **Revolved Boss/Base** as shown in Figure 36.

Figure 36

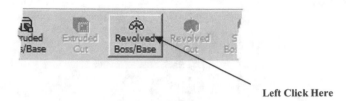

Left Click Here

46. Move the cursor over the "axis" line that was drawn above the part as shown in Figure 37. The line will turn red.

Figure 37   Left Click Here

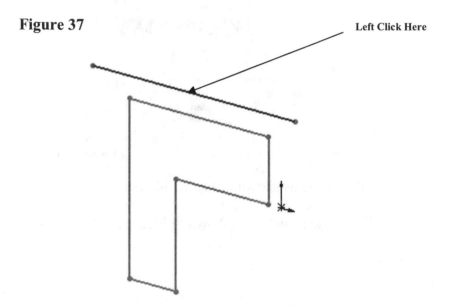

47. Left click once on the line as shown in Figure 37.

48. Move the cursor inside the profile causing it to turn red. Left click once as shown in Figure 38.

Figure 38

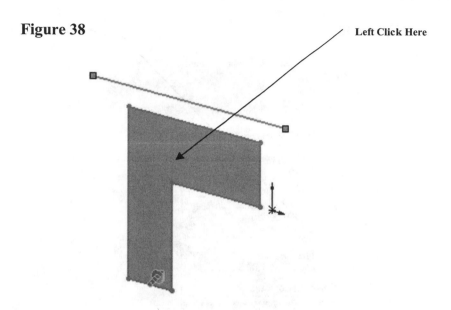

Left Click Here

49. A preview of the revolve will appear as shown in Figure 39.

Figure 39

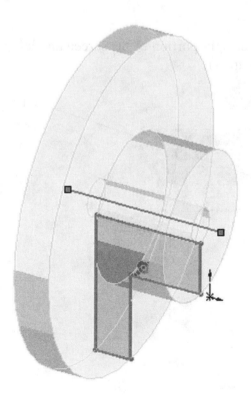

50.   Move the cursor to the middle left portion of the screen and left click on the checkmark next to the text "Thin Feature". SolidWorks will remove the checkmark from the box as shown in Figure 40.

Figure 40

Left Click Here

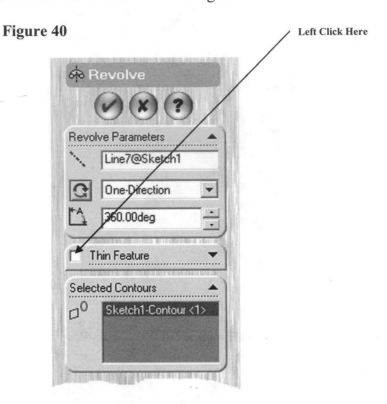

51.   Move the cursor to the upper right portion of the screen and left click on the green checkmark as shown in Figure 41.

Figure 41

Left Click Here

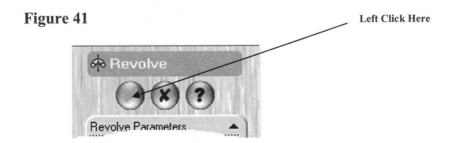

52.   Your screen should look similar to Figure 42. You may have to use the zoom out command to view the entire part.

Figure 42

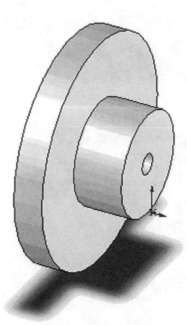

53.   Move the cursor to the edge of the part causing the edges to turn red. After the edges become red, right click on the surface as shown in Figure 43.

Figure 43   Right Click Here

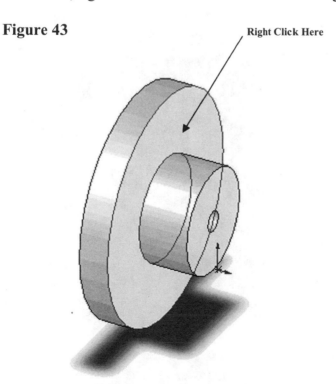

97

54.   The surface will turn green and a pop up menu will appear. Left click on **Insert Sketch** as shown in Figure 44.

Figure 44

Left Click Here

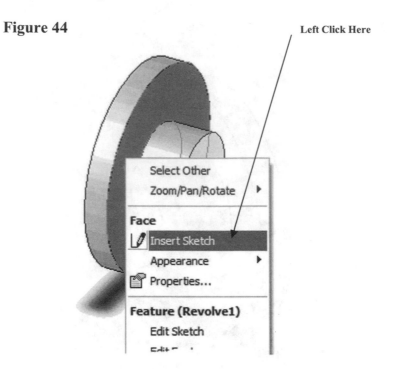

55.   SolidWorks will begin a new sketch on the selected surface. Your screen should look similar to Figure 45.

Figure 45

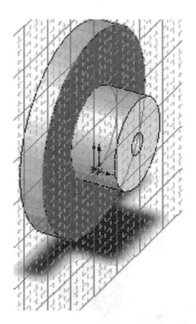

56. Move the cursor to the upper right portion of the screen and left click on the drop down arrow next to the "Standard Views" icon. A drop down menu will appear. Left click on **Right** as shown in Figure 46.

Figure 46

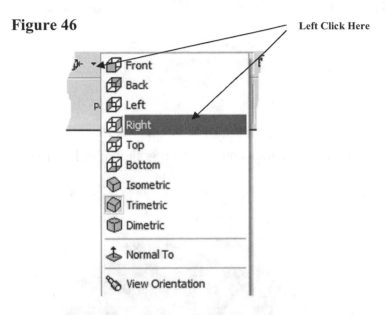

Left Click Here

57. SolidWorks will rotate the part to provide a perpendicular view of the selected surface as shown in Figure 47.

Figure 47

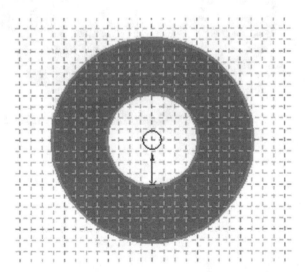

58.   Move the cursor to the upper left corner of the screen and left click on **Line** as shown in Figure 48.

Figure 48

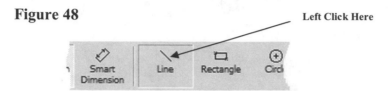

Left Click Here

59.   Move the cursor over the small circle and wait a few seconds. A center marker will appear. Left click on the center marker as shown in Figure 49.

Figure 49

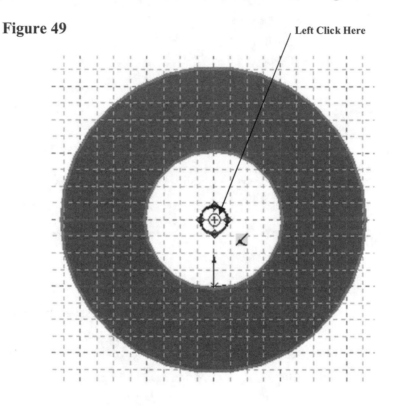

Left Click Here

60. Move the cursor straight up and left click as shown in Figure 50.

Figure 50

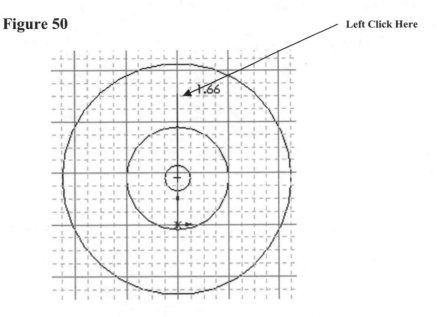

61. Now right click. A pop up menu will appear. Left click on **Select** as shown in Figure 51.

Figure 51

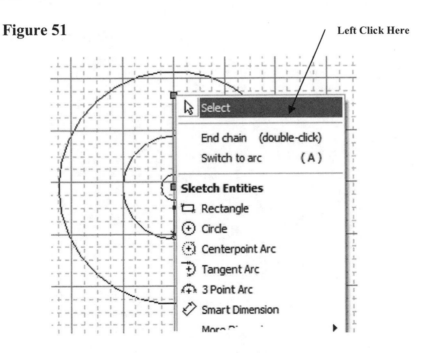

62.   Move the cursor to the upper left portion of the screen and left click on
        Smart Dimension as shown in Figure 52.

        Figure 52

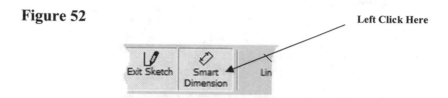

63.   Move the cursor over the line that was just drawn. The line will turn red as shown
        in Figure 53. Select the line by left clicking anywhere on the line **or** on each of
        the end points. The dimension will be attached to the cursor.

        Figure 53

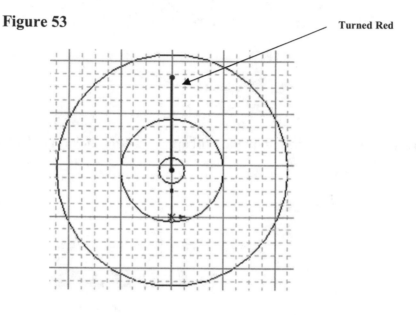

64. Move the cursor to where the dimension will be placed and left click once as shown in Figure 54.

Figure 54

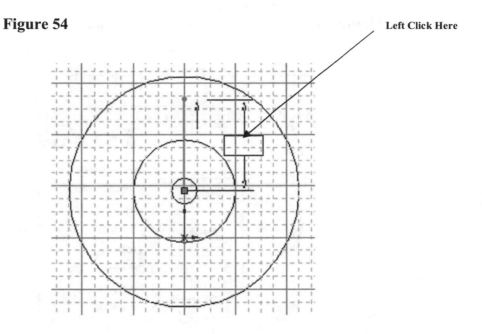

65. The Modify dialog box will appear as shown in Figure 55.

Figure 55

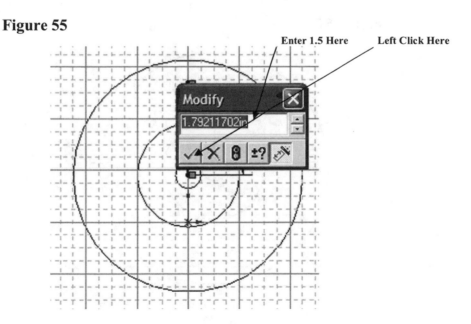

66. To edit the dimension, type **1.5** in the Modify dialog box (while the current dimension is highlighted) as shown in Figure 55. Left click on the green checkmark or press **Enter** on the keyboard.

67.   Right click anywhere on the screen. A pop up menu will appear. Left click on **Select** as shown in Figure 56.

Figure 56

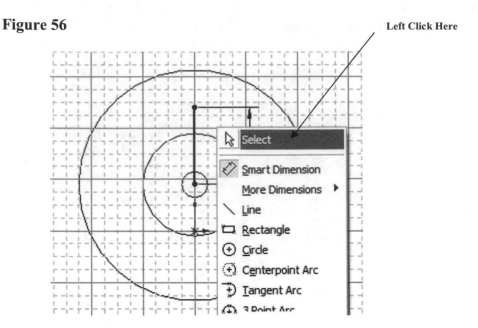

68.   The dimension of the line will become 1.5 inches as shown in Figure 55. Use the Zoom icons to zoom out if necessary.

Figure 57

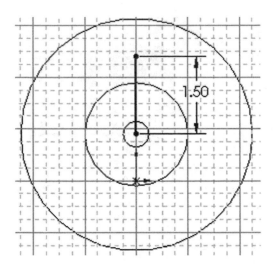

69. Move the cursor to the upper left portion of the screen and left click on **Circle** as shown in Figure 58.

Figure 58

Left Click Here

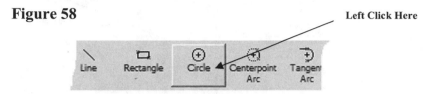

70. Left click on the endpoint of the line as shown in Figure 59.

Figure 59

Left Click Here

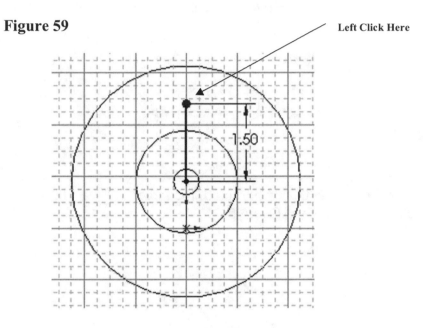

71. Move the cursor out to create a circle as shown in Figure 60.

Figure 60

Move Cursor Here

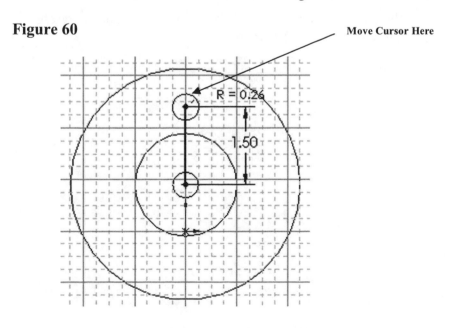

72.　　Left click as shown in Figure 61.

Figure 61

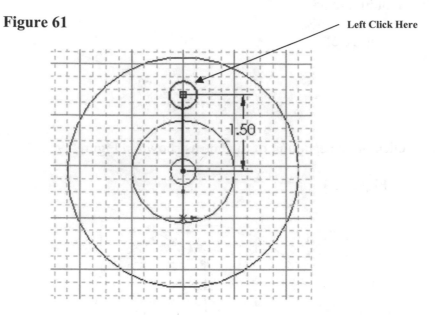

73.　　Move the cursor to the left portion of the screen and left click on the green checkmark as shown in Figure 62.

Figure 62

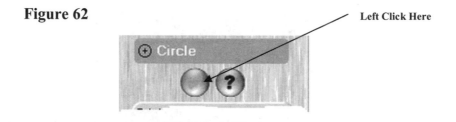

74.　　Move the cursor to the middle left portion of the screen and left click on **Smart Dimension** as shown in Figure 63.

Figure 63

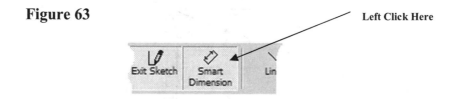

75. Move the cursor to the circle that was just drawn. The circle will turn red. Select the circle by left clicking anywhere on the circle (not the center) as shown in Figure 64. The dimension will be attached to the cursor.

Figure 64

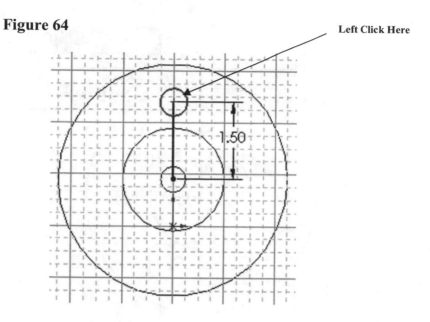

76. Move the cursor to where the dimension will be placed and left click once as shown in Figure 65.

Figure 65

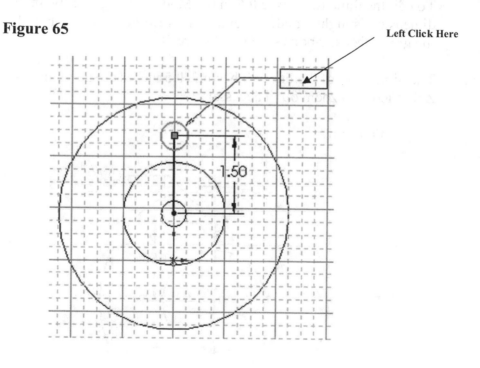

77. The Modify dialog box will appear as shown in Figure 66.

Figure 66

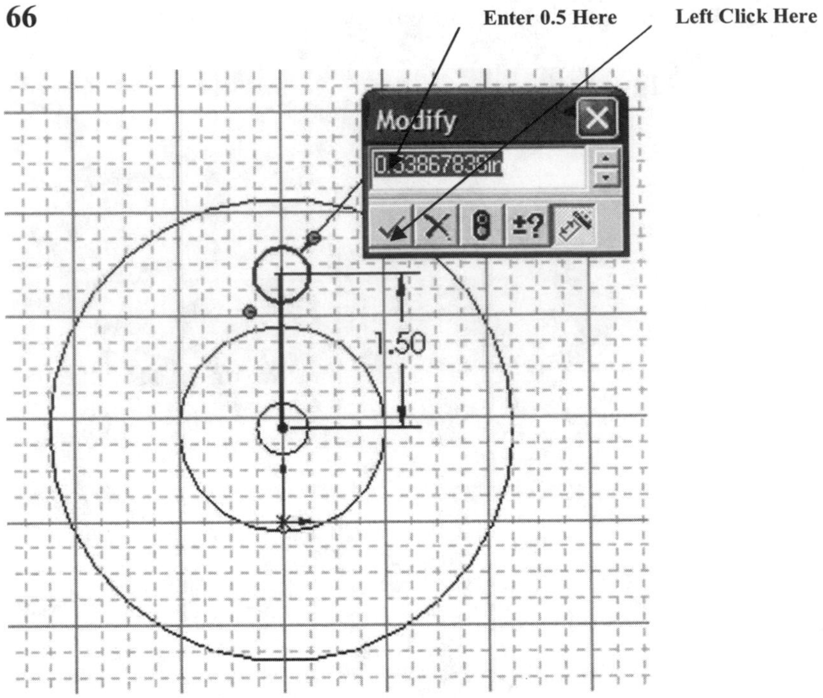

Enter 0.5 Here   Left Click Here

78. To edit the dimension, type **0.5** in the Modify dialog box (while the current dimension is highlighted) and either press **Enter** on the keyboard or left click on the green checkmark as shown in Figure 66.

79. The dimension of the line will become **0.5** inches as shown in Figure 67. Use the Zoom icons to zoom out if necessary.

Figure 67

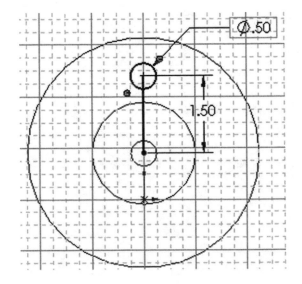

80.   Right click anywhere on the screen. A pop up menu will appear. Left click on **Select** as shown in Figure 68.

Figure 68

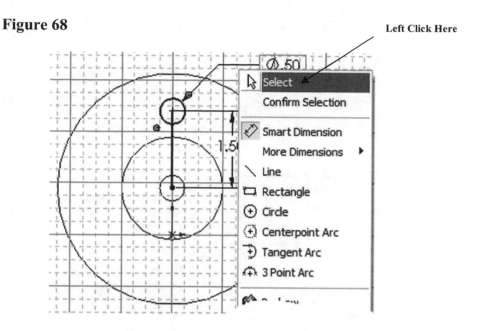

81.   Move the cursor to the line that was used to locate the center of the circle. The line will turn red as shown in Figure 69.

Figure 69

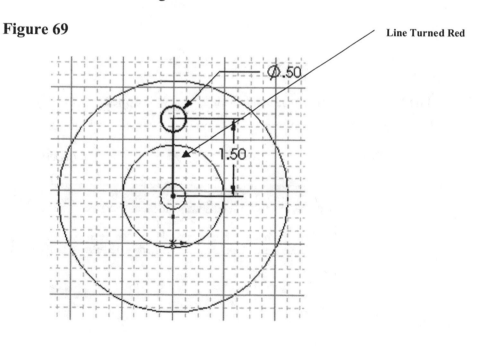

82. Right click on the line after it turns red. A pop up menu will appear as shown in Figure 70.

Figure 70

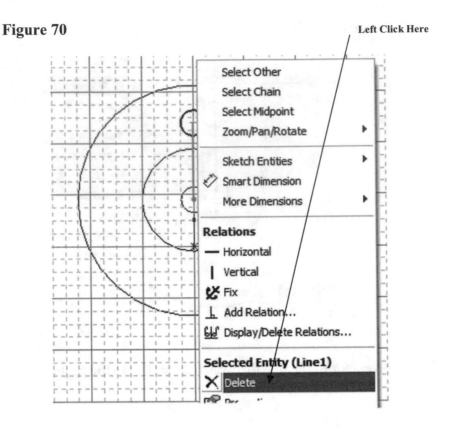

83. Left click on **Delete** as shown in Figure 70.

84. The Sketcher Confirm Delete dialog box will appear. Left click on **Yes** as shown in Figure 71.

Figure 71

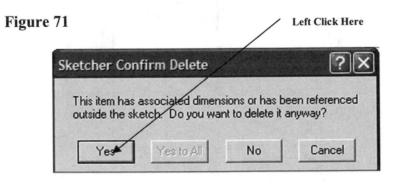

85. After you have verified that no commands are active, right click anywhere on the sketch. A pop up menu will appear. Left click on **Exit Sketch** as shown in Figure 72.

Figure 72

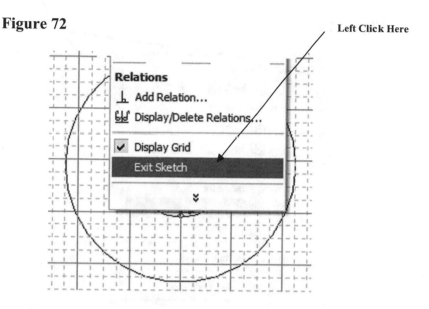

86. SolidWorks is now out of the Sketch commands and into the Features commands. Notice that the commands at the top of the screen are now different. To work in the Features commands a sketch must be present and have no opens (non-connected lines). If there are any opens in the sketch an error message will appear. Your screen should look similar to Figure 73.

Figure 73

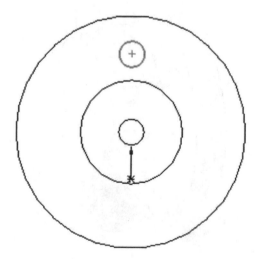

87. Move the cursor to the upper right portion of the screen and left click on the drop down arrow next to the "Standard Views" icon. A drop down menu will appear. Left click on **Trimetric** as shown in Figure 74.

Figure 74

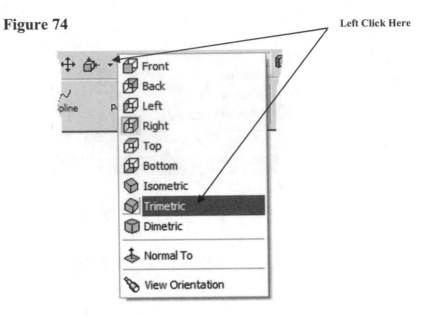

88. The view will become trimetric as shown in Figure 75.

Figure 75

89. Move the cursor to the upper middle portion of the screen and left click on **Extruded Cut** as shown in Figure 76.

Figure 76

Left Click Here

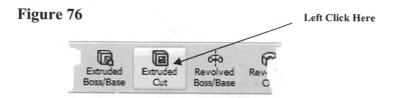

90. Move the cursor to the edge of the circle. The edge will turn red as shown in Figure 77.

Figure 77

Turned Red

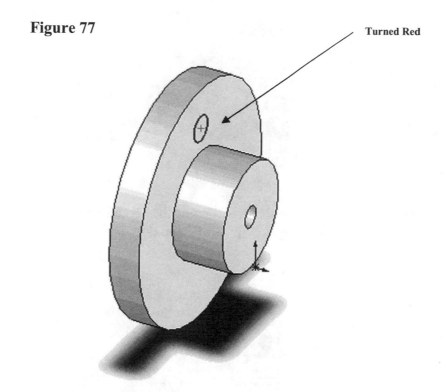

91. After the hole turns red, left click once.

92.   A preview of the hole is displayed as shown in Figure 78.

Figure 78

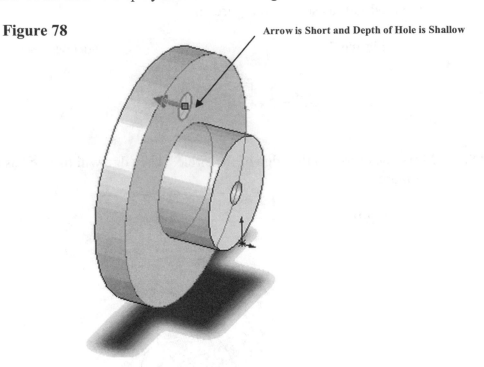

Arrow is Short and Depth of Hole is Shallow

93.   Move the cursor to the middle left portion of the screen. Highlight the text located next to D1 (as shown in Figure 79) and enter **.75**.

Figure 79

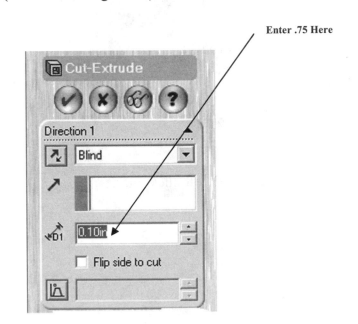

Enter .75 Here

94.   Notice that the depth of the hole has changed in the preview. The arrow
        illustrating the hole depth is now longer as shown in Figure 80.

Figure 80

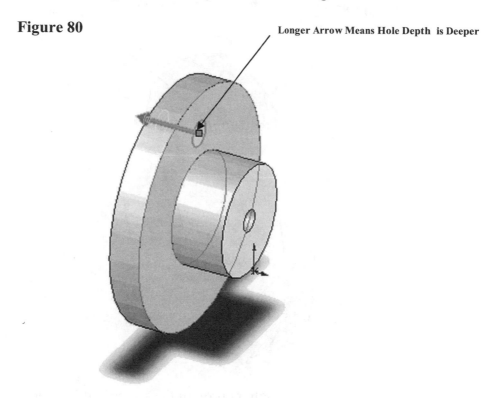

Longer Arrow Means Hole Depth is Deeper

95.   Left click on the green checkmark as shown in Figure 81.

Figure 81

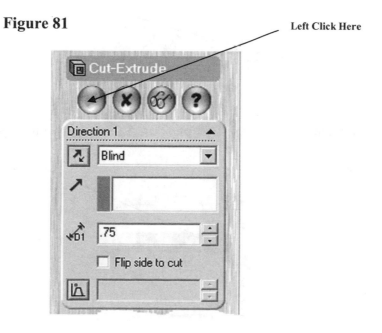

Left Click Here

96. Your screen should look similar to Figure 82.

Figure 82

Thru Hole

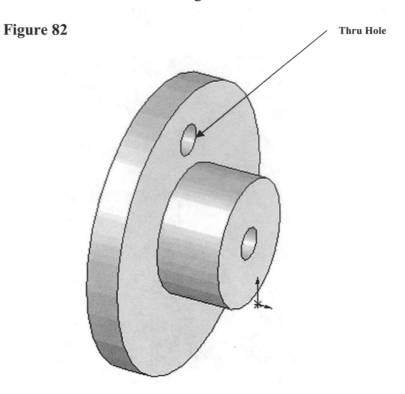

97. Move the cursor to the far right portion of the screen and left click on **Reference Geometry** as shown in Figure 83.

Figure 83

Left Click Here

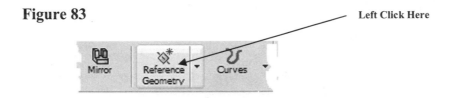

98. A drop down menu will appear. Left click on **Axis** as shown in Figure 84.

Figure 84

Left Click Here

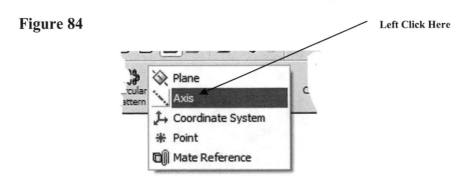

99.   Move the cursor over the center of the center hole causing it to turn red as shown in Figure 85.

Figure 85

Turned Red

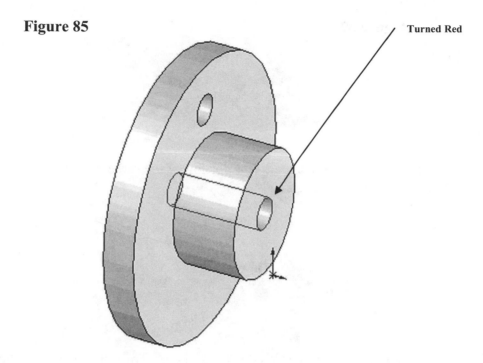

100.   After the inside of the hole turns red, left click once. A center axis will be placed inside the hole as shown in Figure 86.

Figure 86

Center Axis

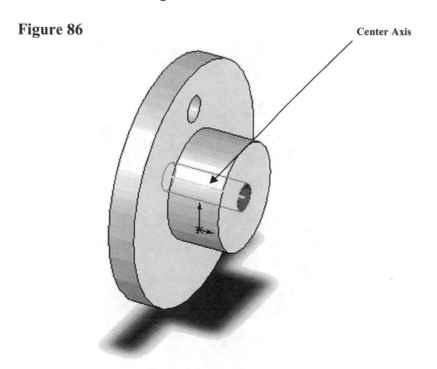

101.   Move the cursor to the middle left portion of the screen and left click on the green checkmark as shown in Figure 87.

Figure 87   Left Click Here

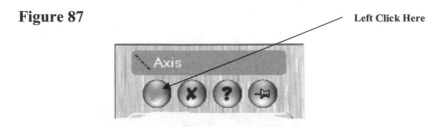

102.   Move the cursor to the upper right portion of the screen and left click on **Circular Pattern** as shown in Figure 88.

Figure 88   Left Click Here

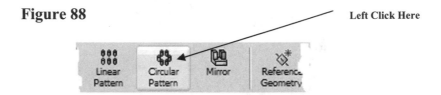

103.   Move the cursor to the inside edge of the hole. The outside edge of the entire hole will turn red. Left click once as shown in Figure 89.

Figure 89   Left Click Here

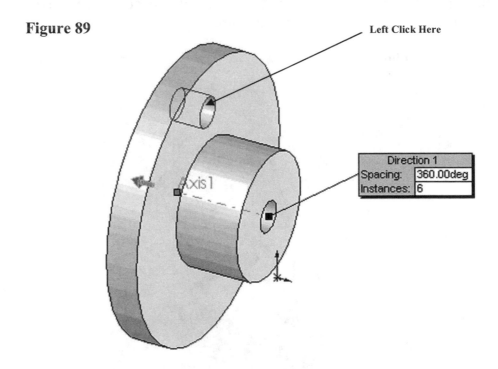

104. Move the cursor to the left portion of the screen. Under "Parameters" place a checkmark in the box next to the text "Equal Spacing". The degree amount automatically changes to 360 degrees. Enter **6** for the number of holes directly below 360 as shown in Figure 90.

Figure 90

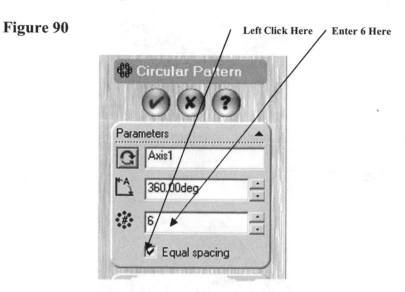

Left Click Here   Enter 6 Here

105. A preview of the circular pattern will be displayed as shown in Figure 91.

Figure 91

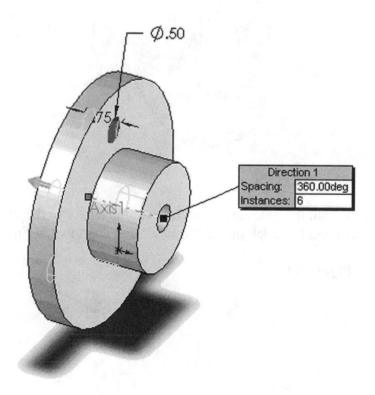

106. Move the cursor to the upper left portion of the screen and left click on the green checkmark as shown in Figure 92.

Figure 92

Left Click Here

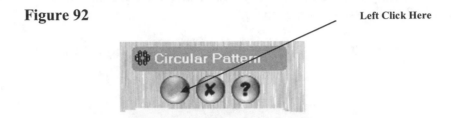

107. Your screen should look similar to Figure 93.

Figure 93

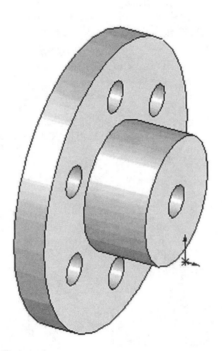

108. To ensure that the holes are correct, move the cursor to the upper right portion of the screen and left click on the "Rotate" icon as shown in Figure 94.

Figure 94

Left Click Here

109. The "Rotate" command will become active. Left click anywhere on the screen. Hold the left mouse button down and drag the cursor around. The part will rotate as shown in Figure 95.

Figure 95

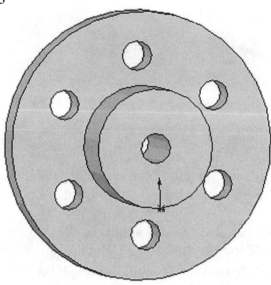

110. Right click around the part. A pop up menu will appear. Left click on **Select**. Other viewing options are shown in the pop up menu in Figure 96.

Figure 96

Left Click Here

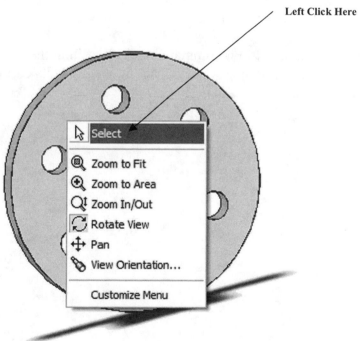

111. The same commands for viewing are located at the top of the screen as shown in Figure 97.

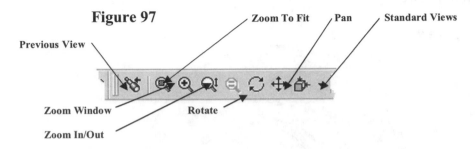

Figure 97

Previous View

Zoom To Fit   Pan   Standard Views

Zoom Window   Rotate

Zoom In/Out

112. The Zoom Window command works by using the cursor to draw a window around the area you want to zoom in on. After selecting the "Zoom Window" icon hold the left mouse button down and drag a diagonal box around the desired area. Release the left mouse button when the proper amount of zoom is achieved.

113. The Zoom In/Out command works similar to the Zoom Window command. Start by selecting the "Zoom In/Out" icon. Left click on the drawing and hold the left mouse button down while moving the cursor up and down until the proper amount of zoom is achieved.

114. The Pan command works similar to the Zoom In/Out command. Start by selecting the "Pan" icon. Left click on the drawing and hold the left mouse button down while moving the cursor up and down or side to side. Release the mouse button after the desired view is achieved.

115. The Zoom to Fit command works by filling the screen so that the entire part is as large as possible while remaining visible in its entirety.

Drawing Activities

Problem 1

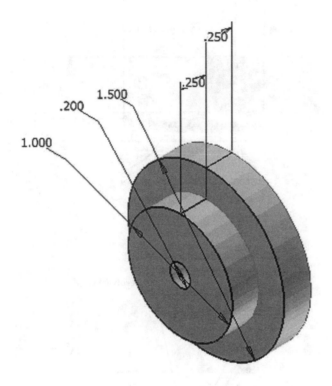

Problem 2

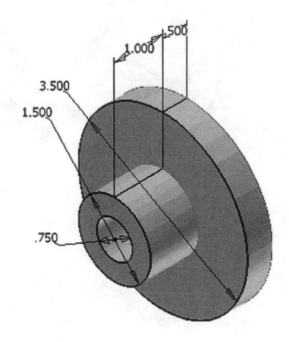

Problem 3

Revolve Axis

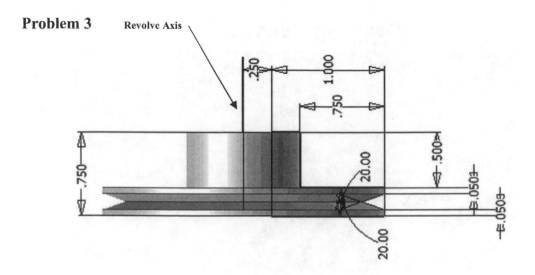

Problem 4

Revolve Axis

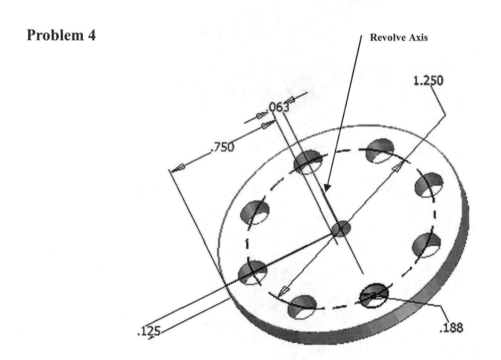

1.250

.063

.750

.125

.188

124

Problem 5

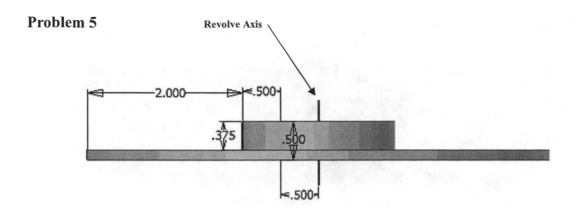

Revolve Axis

2.000 .500 .375 .500 .500

Problem 6

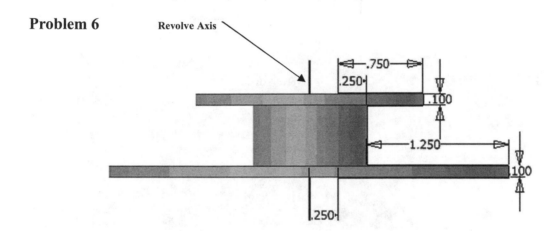

Revolve Axis

.750 .250 .100 1.250 .100 .250

Problem 7

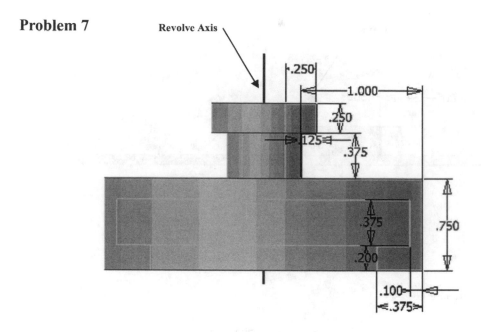

Problem 8

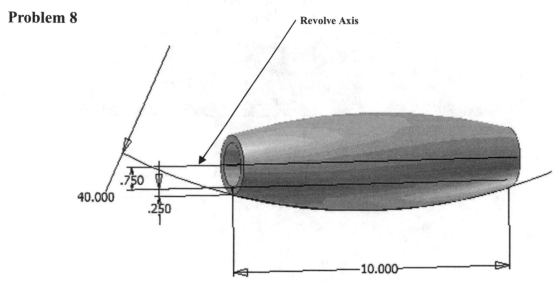

Chapter 3 Learning To Create a Detail Drawing

Objectives:

- Create a simple sketch using the Sketch commands
- Extrude a sketch into a solid using the Features commands
- Create an Orthographic view using the Drawing commands
- Create a Solid Model view using the Edit Views command

Chapter 3 includes instruction on how to design the parts shown below.

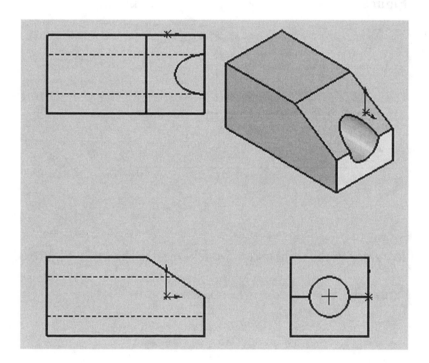

1.   Start SolidWorks by referring to "Chapter 1 Getting Started".

2.   After SolidWorks is running, begin a new sketch.

3.   Move the cursor to the upper left corner of the screen and left click on **Line** as shown in Figure 1.

Figure 1

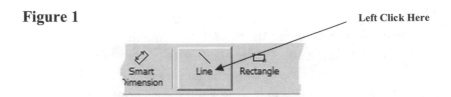

4.   Move the cursor somewhere in the lower left portion of the screen and left click once. This will be the beginning end point of a line as shown in Figure 2.

Figure 2

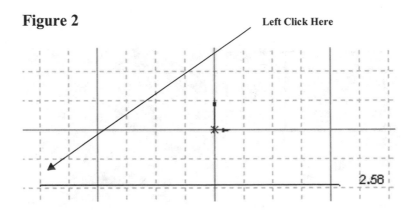

5.   Move the cursor to the right and left click once as shown in Figure 3.

Figure 3

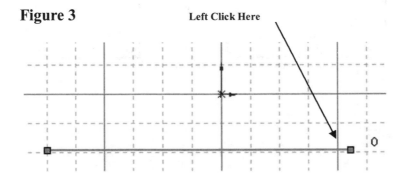

6.   Move the cursor upward and left click once as shown in Figure 4.

Figure 4

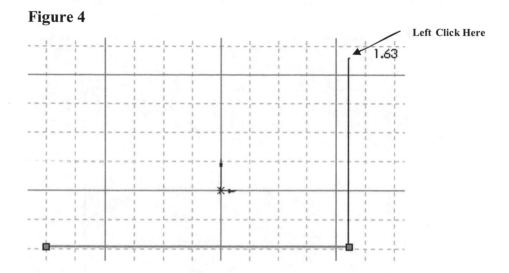

7.   Move the cursor to the left and wait for dots to appear. Left click once as shown in Figure 5.

Figure 5

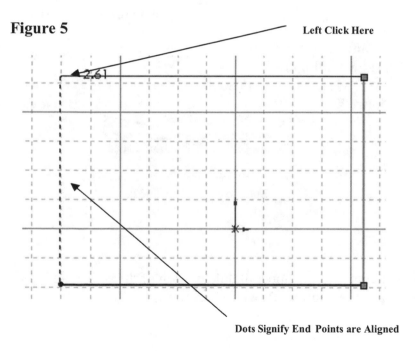

129

8.   Move the cursor back to the original starting end point. A red dot will appear. Left click once. Your screen should look similar to Figure 6.

Figure 6

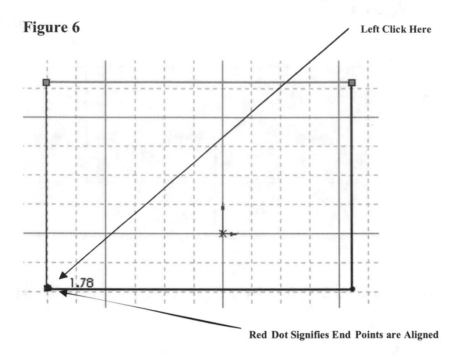

9.   Right click anywhere on the screen. A pop up menu will appear. Left click on **Select** as shown in Figure 7.

Figure 7

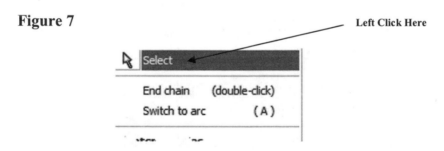

10. Move the cursor to the upper middle portion of the screen and left click on **Smart Dimension** as shown in Figure 8.

Figure 8

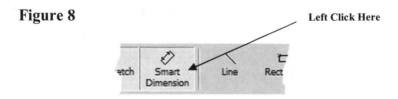

11.   After selecting **Smart Dimension** move the cursor over the bottom horizontal line until it turns red as shown in Figure 9. Select the line by left clicking anywhere on the line **or** on each of the end points. The dimension will be attached to the cursor.

Figure 9

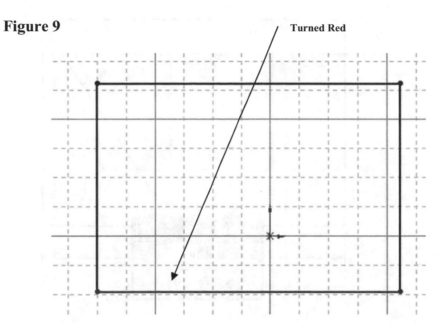

Turned Red

12.   Move the cursor to where the dimension will be placed and left click once as shown in Figure 10.

Figure 10

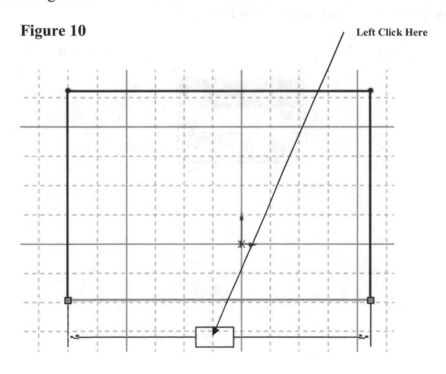

Left Click Here

131

13. The Modify dialog box will appear as shown in Figure 11.

Figure 11

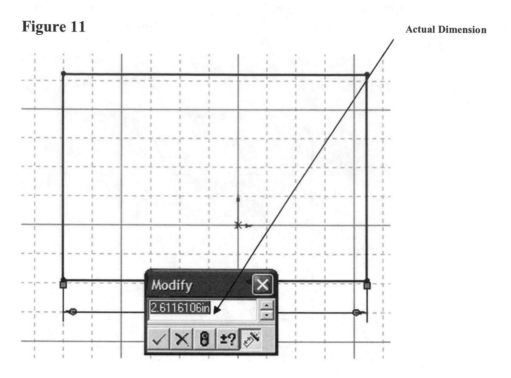

14. To edit the dimension, type **2.00** in the Modify dialog box (while the current dimension is highlighted) and either press **Enter** on the keyboard or left click on the green checkmark as shown in Figure 12.

Figure 12

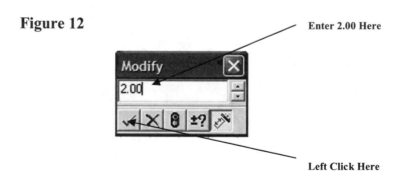

15.   The dimension of the line will become 2.00 inches as shown in Figure 13.

Figure 13

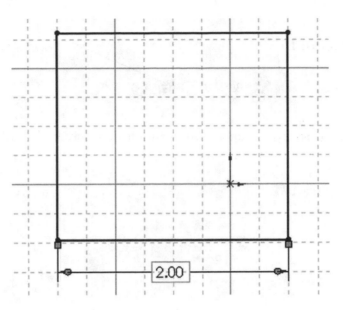

16.   Right click anywhere on the screen. A pop up menu will appear. Left click on **Select** as shown in Figure 14.

Figure 14   Left Click Here

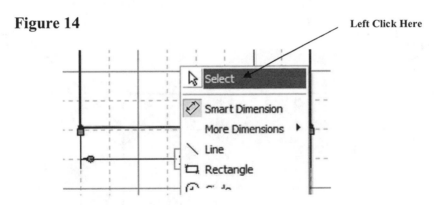

17.   To view the entire drawing, move the cursor to the upper right portion of the screen and left click once on the "Zoom to Fit" icon as shown in Figure 15.

Figure 15   Left Click Here

18. The drawing will "fill up" the entire screen. If the drawing is still too large, left click on the "Zoom In/Out" icon as shown in Figure 16. Hold the left mouse button down and move the cursor up and down to achieve the desired view of the sketch.

Figure 16   **Left Click Here**

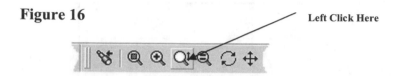

19. Move the cursor to the upper left portion of the screen and left click on **Smart Dimension** as shown in Figure 17.

Figure 17   **Left Click Once**

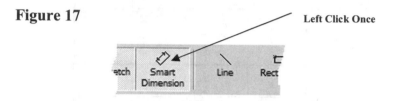

20. Move the cursor over the right side line until it turns red as shown in Figure 18. Select the line by left clicking anywhere on the line **or** on each of the end points. The dimension will be attached to the cursor.

Figure 18   **Turned Red**

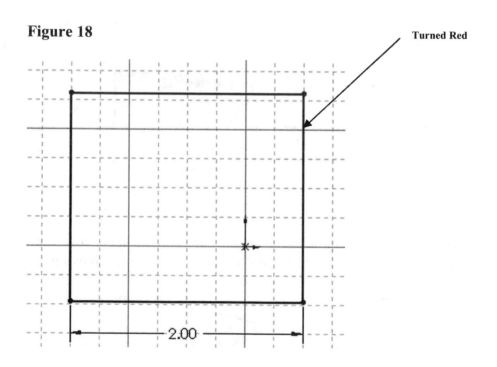

21. Move the cursor to where the dimension will be placed and left click once as shown in Figure 19.

Figure 19

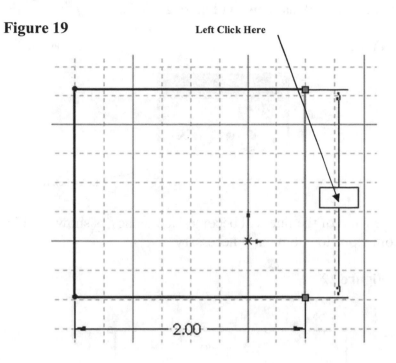

Left Click Here

2.00

22. The Modify dialog box will appear as shown in Figure 20.

Figure 20

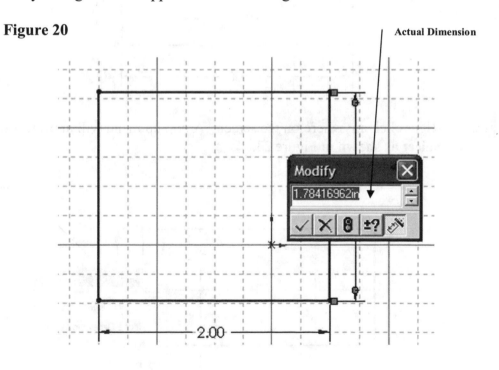

Actual Dimension

Modify

1.78416962in

2.00

135

23. To edit the dimension, type **1.00** in the Modify dialog box (while the current dimension is highlighted) and either press **Enter** on the keyboard or left click on the green checkmark as shown in Figure 21.

Figure 21

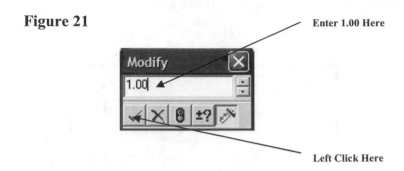

Enter 1.00 Here

Left Click Here

24. The dimension of the line will become 1.00 inches as shown in Figure 22. Use the Zoom icons to zoom out if necessary.

Figure 22

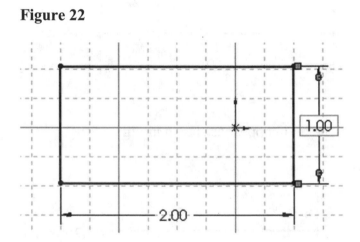

25. Right click anywhere on the screen. A pop up menu will appear. Left click on **Select** as shown in Figure 23.

Figure 23

Left Click Here

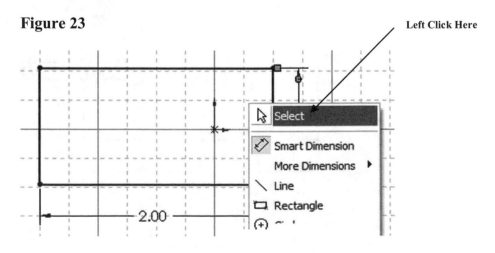

26.   Move the cursor to the upper left portion of the screen and left click on **Smart Dimension** as shown in Figure 24.

Figure 24

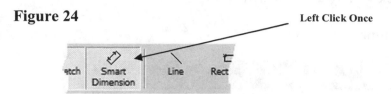

27.   After selecting **Smart Dimension** move the cursor over the top horizontal line until it turns red as shown in Figure 25. Select the line by left clicking anywhere on the line <u>or</u> on each of the end points. This will cause the dimension to be attached to the cursor.

Figure 25

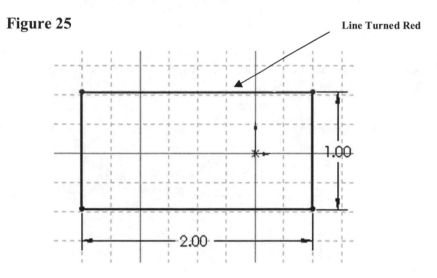

28.   Move the cursor to where the dimension will be placed and left click once as shown in Figure 26.

Figure 26

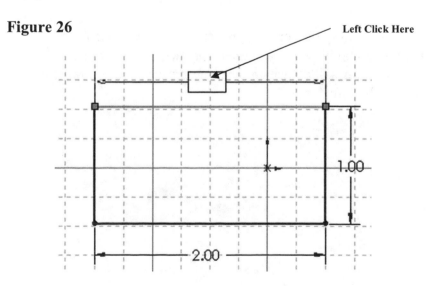

29.   Notice that the dimension is exactly 2.000. Move the cursor to where the dimension will be placed and left click once. The Make Dimension Driven? dialog box will appear as shown in Figure 27.

Figure 27

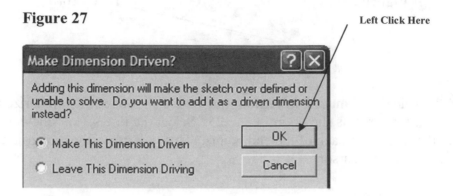

30.   Right click anywhere on the screen. A pop up menu will appear. Left click on **Select** as shown in Figure 28.

Figure 28

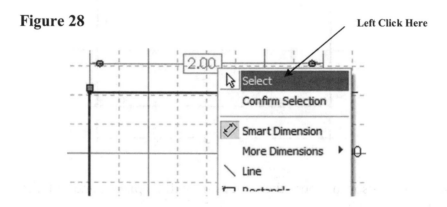

31.   This dimension will over constrain the sketch because the sketch has been constrained with 90 degree angles when it was constructed. Left click on **OK**. The dimension will be driven meaning it cannot be used to edit or change the length of the line.

138

32. The driven dimension appears grayed out as shown in Figure 29.

Figure 29

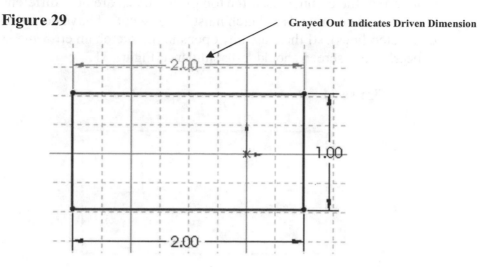

Grayed Out Indicates Driven Dimension

33. Dimensioning the far left line would also result in a driven dimension. Because of this, the dimensioning portion is complete.

34. After the sketch is complete it is time to extrude the sketch into a solid.

35. After you have verified that no commands are active, right click anywhere on the sketch. A pop up menu will appear. Left click on **Exit Sketch** as shown in Figure 30.

Figure 30

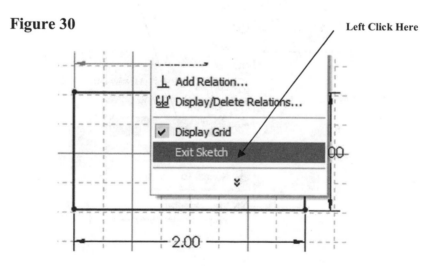

Left Click Here

36. SolidWorks is now out of the Sketch commands and into the Features commands. Notice that the commands at the top of the screen are now different. To work in the Features commands a sketch must be present and have no opens (non-connected lines). If there are any opens in the sketch an error message will appear. Your screen should look similar to Figure 31.

Figure 31

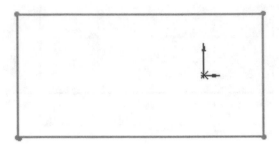

37. Move the cursor to the upper right portion of the screen and left click on the drop down arrow next to the "Standard Views" icon. A drop down menu will appear. Left click on **Trimetric** as shown in Figure 32.

Figure 32   Left Click Here

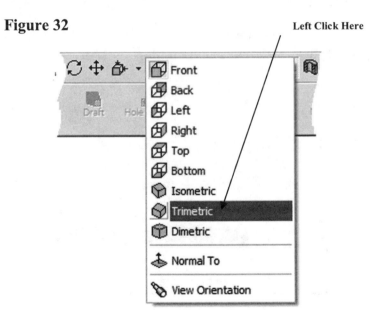

38. The view will become trimetric as shown in Figure 33.

Figure 33

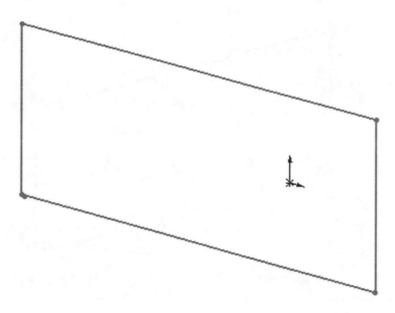

39. Move the cursor to the upper left portion of the screen and left click on **Extruded Boss/Base** as shown in Figure 34. If SolidWorks gave you an error message, there are opens (non-connected lines) somewhere on the sketch. Check each intersection for opens by using the **Extend** and **Trim** commands.

Figure 34   Left Click Here

40.   A preview of the extrusion will be displayed as shown in Figure 35.

Figure 35

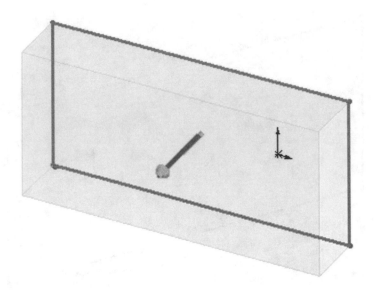

41.   If a preview does not appear, move the cursor over any of the lines. The line will turn red as shown in Figure 36. Left click once.

Figure 36   Turned Red

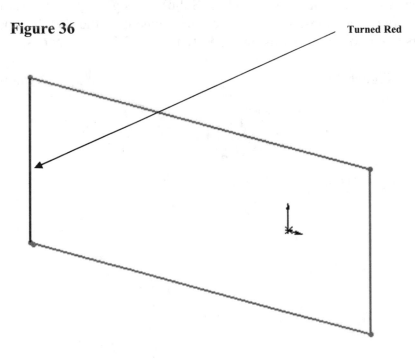

42. Enter **1.00** for D1 as shown in Figure 37.

Figure 37

Enter 1.00 Here

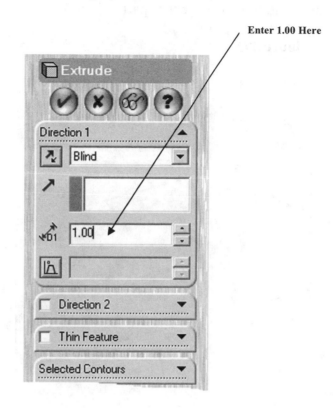

43. Move the cursor to the upper left portion of the screen and left click on the green checkmark as shown in Figure 38.

Figure 38

Left Click Here

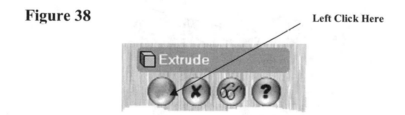

44.   Your screen should look similar to Figure 39. You may have to use the Zoom
Out command to view the entire part.

Figure 39

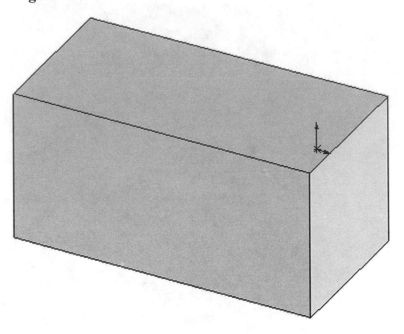

45.   Move the cursor to the upper middle portion of the screen and left click on
Chamfer as shown in Figure 40.

Figure 40   Left Click Here

46.   After selecting **Chamfer,** place a dot (left click) next to the text "Distance distance". Place another dot (left click) next to the text "Full preview" as shown in Figure 41.

Figure 41

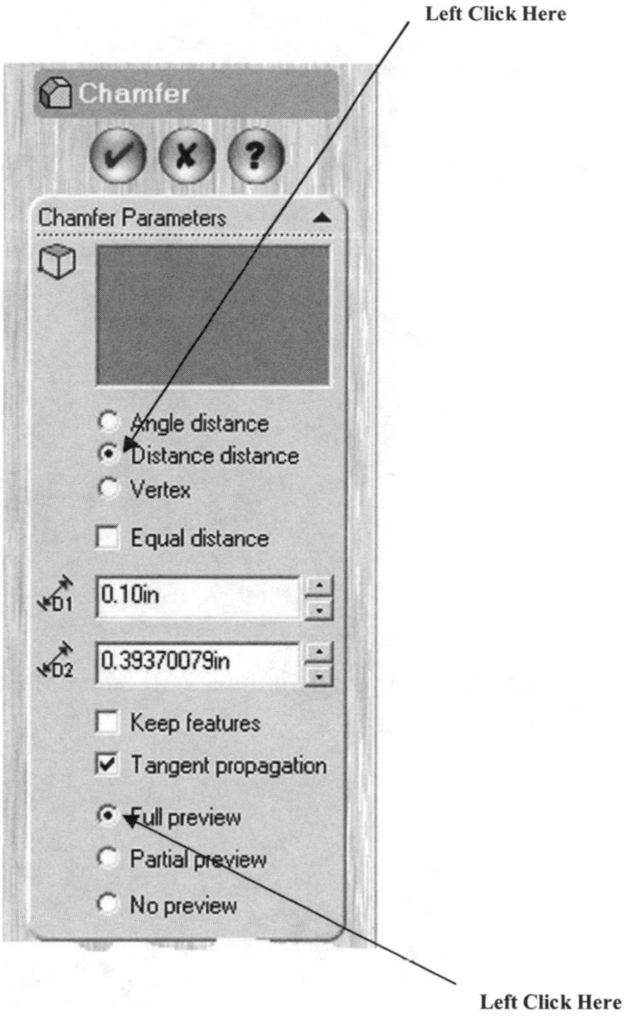

47. Move the cursor to the front upper corner. A red line will appear as shown in Figure 42. Left click once.

Figure 42

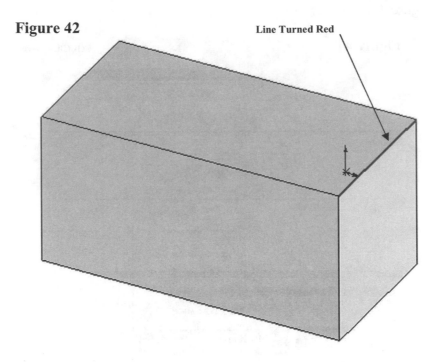

Line Turned Red

48. A preview of the anticipated chamfer will be displayed as shown in Figure 43.

Figure 43

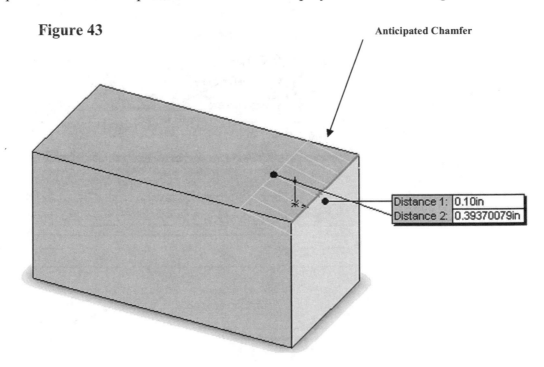

Anticipated Chamfer

Distance 1: 0.10in
Distance 2: 0.39370079in

49.　Move the cursor to D1 in the Chamfer dialog box and highlight the text.　Enter **.50.**　Move the cursor to D2 in the dialog box and highlight the text.　Enter **.75** as shown in Figure 44.

Figure 44

Enter .50 Here　　　Enter .75 Here

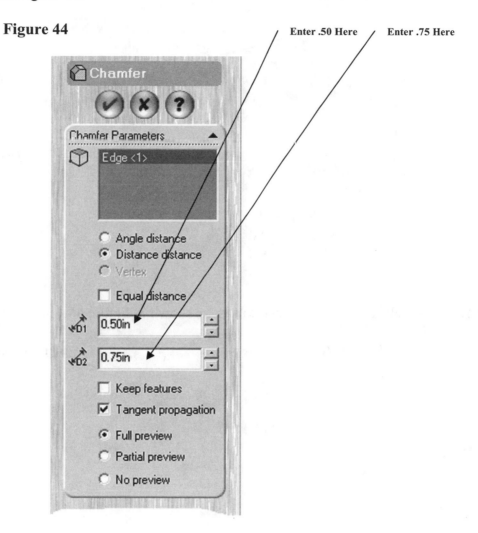

50. A preview of the chamfer will be displayed as shown in Figure 45.

Figure 45

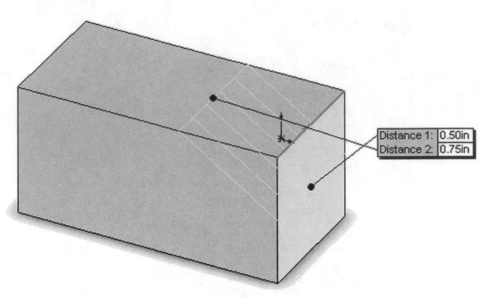

51. Left click on the green checkmark as shown in Figure 46.

Figure 46

Left Click Here

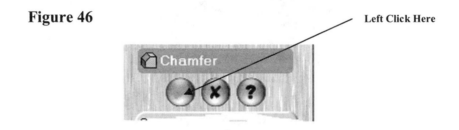

52. Your screen should look similar to Figure 47.

Figure 47

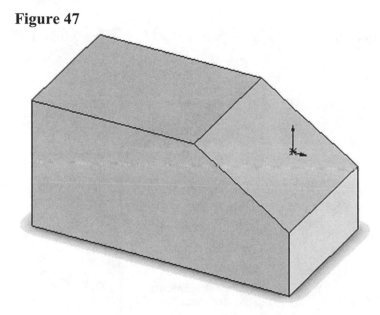

53. Move the cursor to the upper middle portion of the screen and left click on the "Rotate" icon as shown in Figure 48.

Figure 48   **Left Click Here**

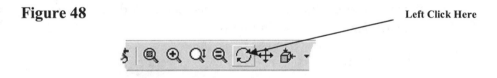

54. While holding the left mouse button down, drag the cursor to the right to gain access to the backside of the part as shown in Figure 49.

Figure 49

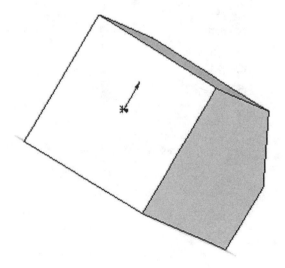

55.   Right click anywhere on the screen. A pop menu will appear. Left click on **Select** as shown in Figure 50.

Figure 50

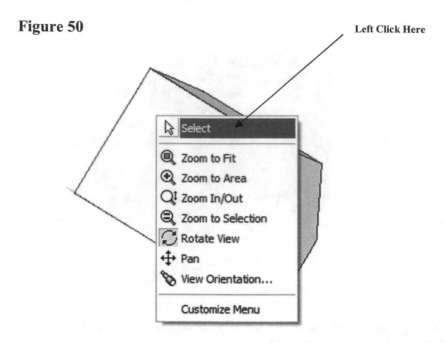

56.   Move the cursor to the upper right portion of the screen and left click on the drop down arrow next to the "Standard Views" icon. A drop down menu will appear. Left click on **Left** as shown in Figure 51.

Figure 51

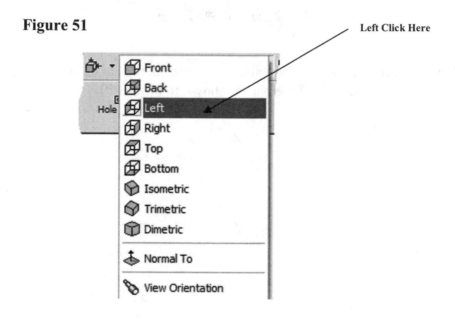

57.	The part will rotate and provide a perpendicular view of the surface as shown in Figure 52.

Figure 52

Right Click Here

58.	Right click anywhere on the surface. A pop up menu will appear. Left click on **Insert Sketch** as shown in Figure 53.

Figure 53

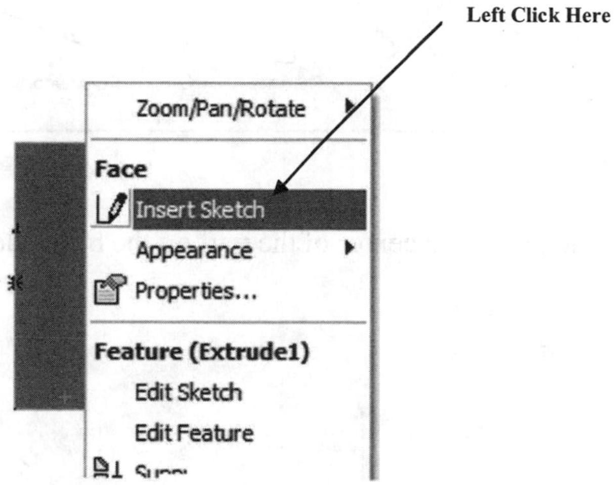

Left Click Here

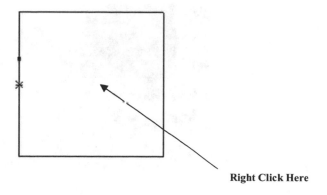

Zoom/Pan/Rotate ▶

Face
Insert Sketch
Appearance ▶
Properties...

Feature (Extrude1)
Edit Sketch
Edit Feature
Supp

151

59.   A new sketch will appear on the surface as shown in Figure 54.

Figure 54

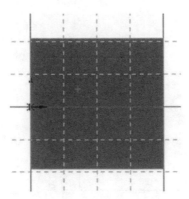

60.   Move the cursor to the upper left portion of the screen and left click on **Circle** as shown in Figure 55.

Figure 55   Left Click Here

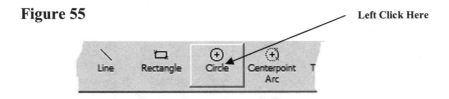

61.   Left click near the center of the part on the backside surface as shown in Figure 56.

Figure 56   Left Click Here

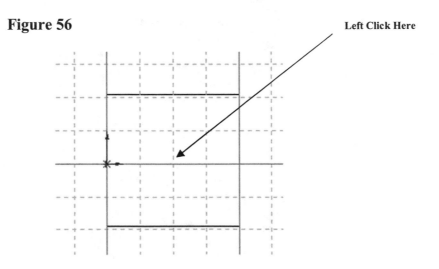

62.   Move the cursor to the side forming a circle and left click as shown in Figure 57.

Figure 57

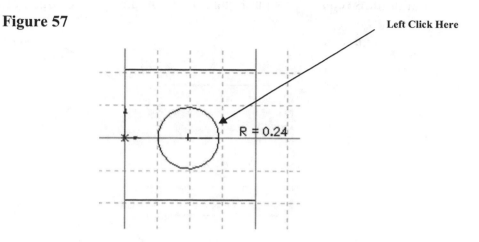

63.   Right click anywhere on the screen. A pop up menu will appear. Left click on **Select** as shown in Figure 58.

Figure 58

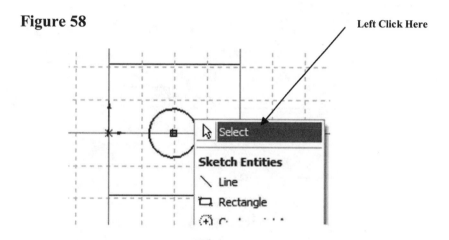

64.   Move the cursor to the upper left portion of the screen and left click on **Smart Dimension** as shown in Figure 59.

Figure 59

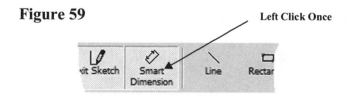

65.   After selecting **Smart Dimension** move the cursor over the edge of the circle until it turns red as shown in Figure 60. Left click on the edge of the circle.

Figure 60

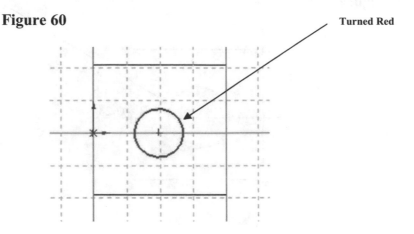

Turned Red

66.   Move the cursor to where the dimension will be placed and left click once as shown in Figure 61.

Figure 61

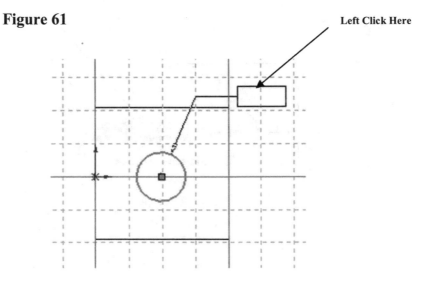

Left Click Here

67.   The Modify dialog box will appear as shown in Figure 62.

Figure 62

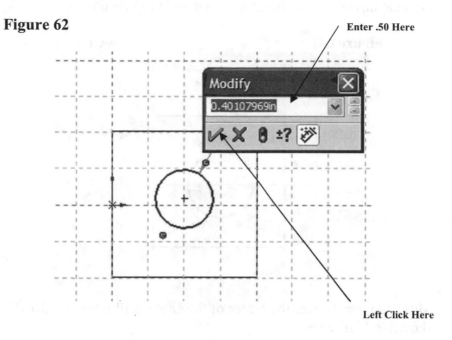

68.   To edit the dimension, type **.50** in the Modify dialog box (while the current dimension is highlighted) and either press **Enter** on the keyboard or left click on the green checkmark as shown in Figure 62.

69.   Your screen should look similar to Figure 63.

Figure 63

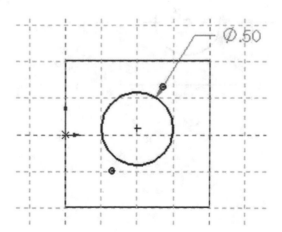

70.   Move the cursor over the edge of the part until it turns red. Select the line by left clicking anywhere on the edge as shown in Figure 64.

Figure 64

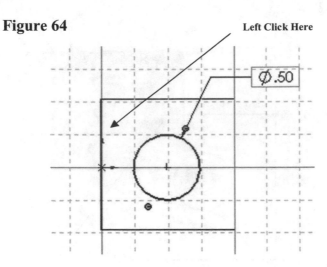

71.   Move the cursor over the center of the circle until it turns red and left click as shown in Figure 65.

Figure 65

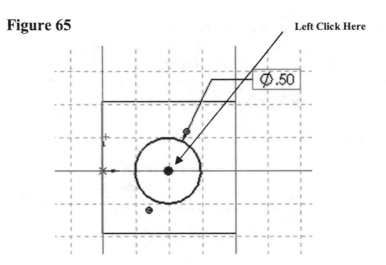

72. Move the cursor to where the dimension will be placed and left click once as shown in Figure 66.

Figure 66

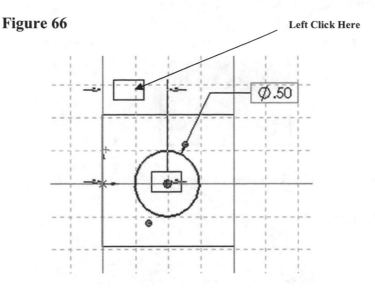

73. The Modify dialog box will appear as shown in Figure 67.

Figure 67

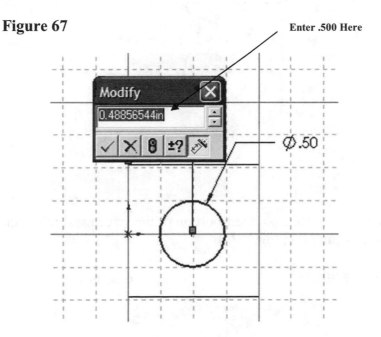

74. To edit the dimension, type **.500** in the Modify dialog box (while the current dimension is highlighted) and either press **Enter** on the keyboard or left click on the green checkmark as shown in Figure 68.

Figure 68

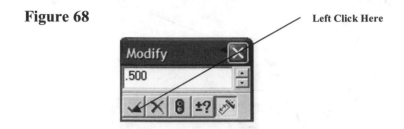

75. Your screen should look similar to Figure 69.

Figure 69

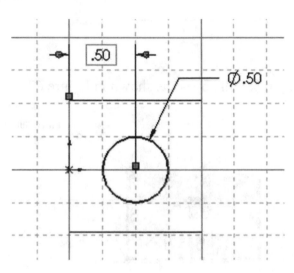

76. Move the cursor to the upper left portion of the screen and left click on **Smart Dimension** as shown in Figure 70.

Figure 70

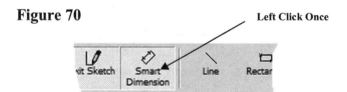

77. Move the cursor to the edge of the part until it turns red. Left click anywhere on the edge as shown in Figure 71.

Figure 71

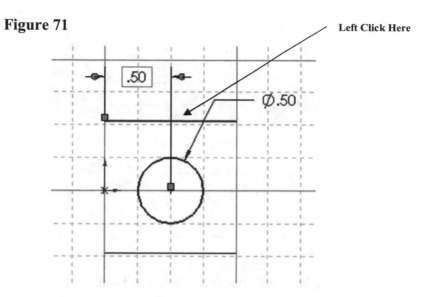

78. Move the cursor over the center of the circle until it turns red and left click as shown in Figure 72.

Figure 72

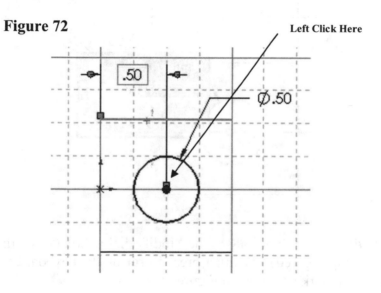

79. Move the cursor to where the dimension will be placed and left click once as shown in Figure 73.

Figure 73

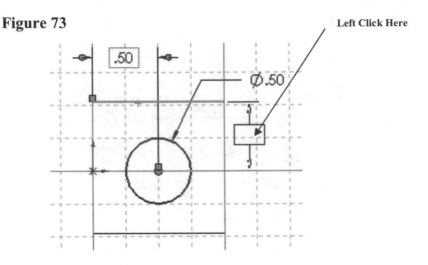

80. The Modify dialog box will appear as shown in Figure 74.

Figure 74

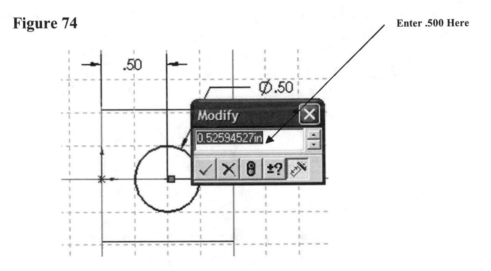

81. To edit the dimension, type **.500** in the Modify dialog box (while the current dimension is highlighted) and either press **Enter** on the keyboard or left click on the green checkmark as shown in Figure 75.

Figure 75

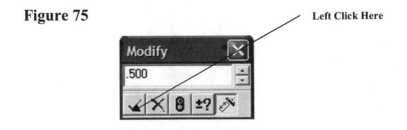

160

82. Your screen should look similar to Figure 76.

Figure 76

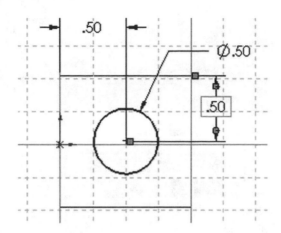

83. After the sketch is complete it is time to extrude a hole through the solid.

84. Right click anywhere on the screen. A pop up menu will appear. Left click on **Select** as shown in Figure 77. This will ensure that no commands are active.

Figure 77

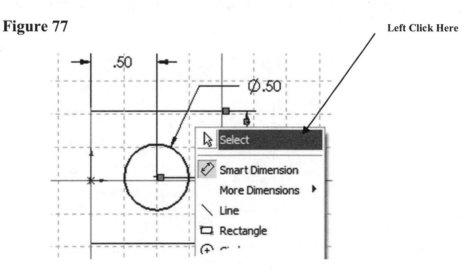

85. After you have verified that no commands are active, right click anywhere on the sketch. A pop up menu will appear. Left click on **Exit Sketch** as shown in Figure 78.

Figure 78

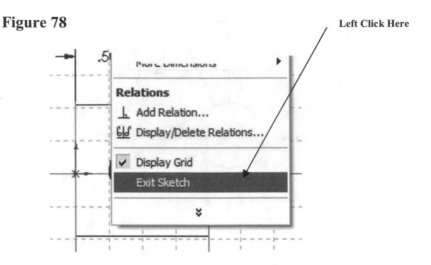

Left Click Here

86. SolidWorks is now out of the Sketch commands and into the Features commands. Notice that the commands at the top of the screen are now different. To work in the Features commands a sketch must be present and have no opens (non-connected lines). If there are any opens in the sketch an error message will appear. Your screen should look similar to Figure 79.

Figure 79

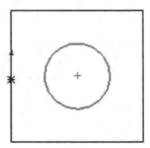

87. Move the cursor to the upper middle portion of the screen and left click on the "Rotate" icon as shown in Figure 80.

Figure 80

Left Click Here

88.   While holding the left mouse button down, drag the cursor to the right to gain an isometric view of the part as shown in Figure 81.

Figure 81

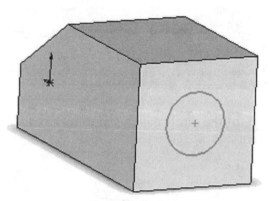

89.   Right click anywhere on the screen. A pop up menu will appear. Left click on **Select** as shown in Figure 82.

Figure 82   **Left Click Here**

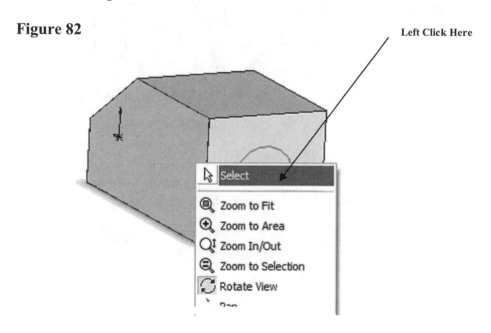

163

90. Move the cursor to the upper middle portion of the screen and left click on **Extruded Cut** as shown in Figure 83. If you received an error message, there are opens (non-connected lines) somewhere on the sketch. Check each intersection for opens by using the **Extend** and **Trim** commands.

Figure 83

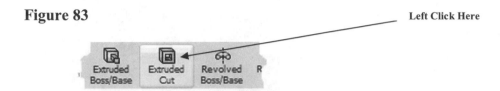

Left Click Here

91. Left click on the edge of the circle causing it to turn red as shown in Figure 84.

Figure 84

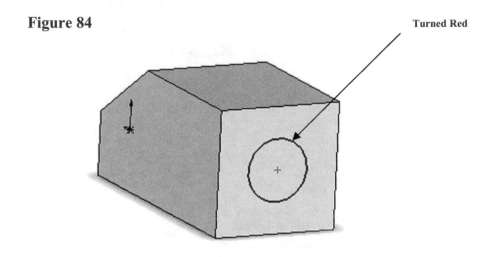

Turned Red

92. Enter **2.00** for D1 as shown in Figure 85.

Figure 85

Enter 2.00 Here

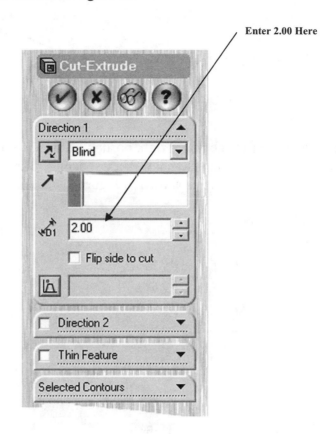

93. A preview will be displayed of the extrusion as shown in Figure 86.

Figure 86

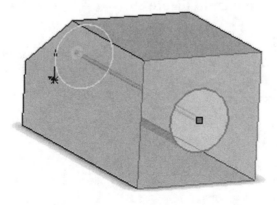

94.   Left click on the green checkmark as shown in Figure 87.

Figure 87

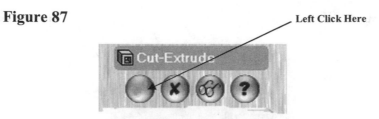

Left Click Here

95.   Move the cursor to the upper right portion of the screen and left click on the drop down arrow next to the "Standard Views" icon. A drop down menu will appear. Left click on **Trimetric** as shown in Figure 88.

Figure 88

Left Click Here

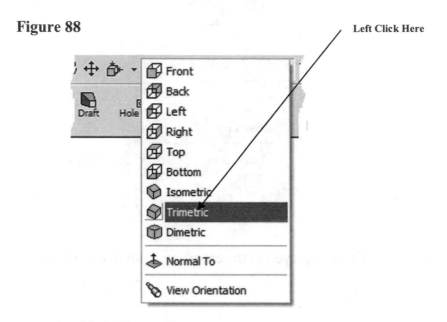

96.   Your screen should similar to Figure 89.

Figure 89

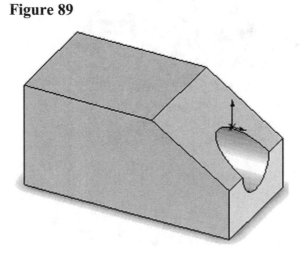

97. Save the part file for easy retrieval to be used in the following section.

98. After the part file has been saved, move the cursor to the upper left portion of the screen and left click on the "Make Drawing" icon as shown in Figure 90.

Figure 90

Left Click Here

99. The Sheet Format/Size dialog box will open as shown in Figure 91.

Figure 91

Left Click Here

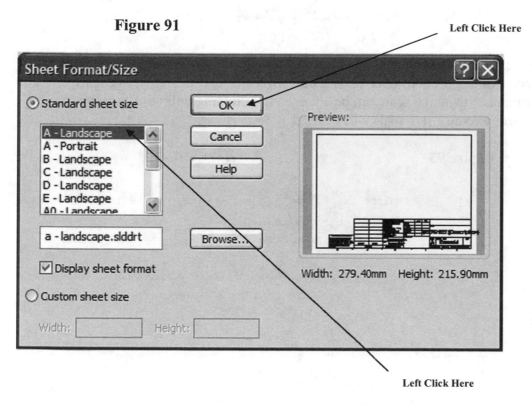

Left Click Here

100. Left click on **A-Landscape** as shown in Figure 91.

101. Left click on **OK**.

102. With the rectangular box attached to the cursor, move the cursor to the lower left portion of the screen under the text "Display Style". Left click on the "Hidden Lines Visible" icon as shown in Figure 92.

Figure 92   Left Click Here

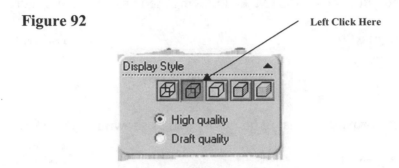

103. Move the cursor to the lower left portion of the screen under the text "Scale" and place a dot (left click) next to "Use custom scale" text as shown in Figure 93. For future drawings, the scale can be changed if a drawing will not fit on the paper. For this exercise, 1:1 scale will be used.

Figure 93   Left Click Here

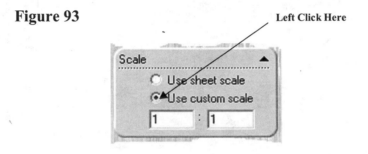

104. Your screen should look similar to Figure 94.

Figure 94

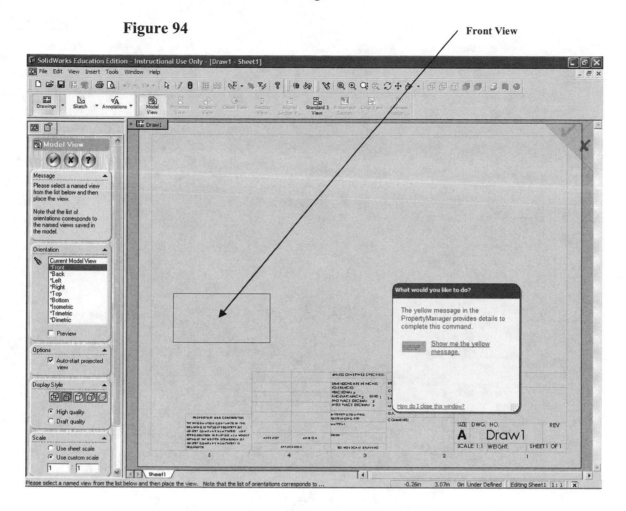

105. SolidWorks is now in the Create Drawing commands. Notice the commands at the left are now different. The rectangular box that is attached to the cursor will become the front view of the part.

106. Move the cursor to the lower left corner of the screen and left click as shown in Figure 95.

Figure 95

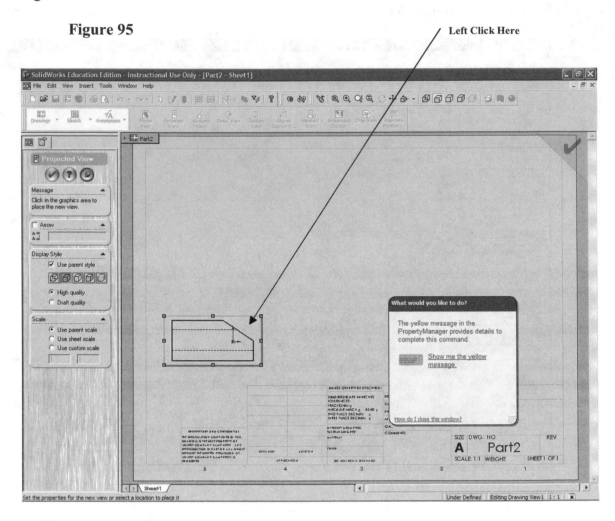

107. A preview of the front view of the part will be displayed as shown in Figure 96.

Figure 96

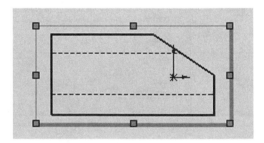

108.   Move the cursor upward. A preview of the top view will be displayed. Left click once as shown in Figure 97.

Figure 97

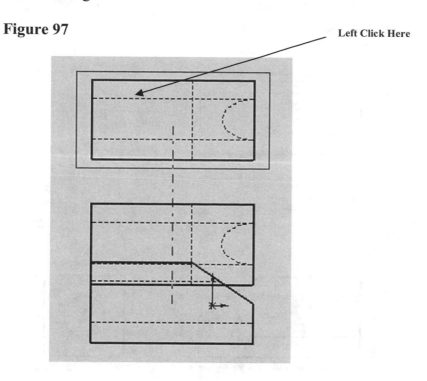

109.   Move the cursor to the right side. A preview of the side view will be displayed. Left click once as shown in Figure 98.

Figure 98

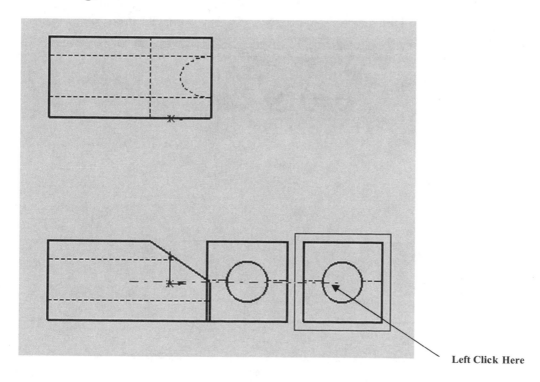

110. After the side view has been placed, right click once. Your screen should look similar to Figure 99.

Figure 99

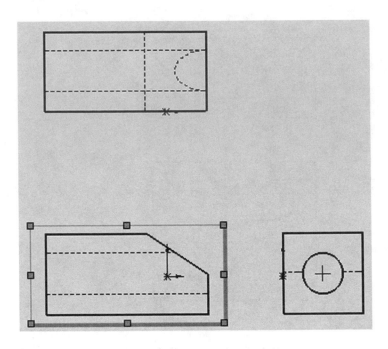

111. Move the cursor to the upper left portion of the screen under Projected View and left click on the green checkmark as shown in Figure 100.

Figure 100

Left Click Here

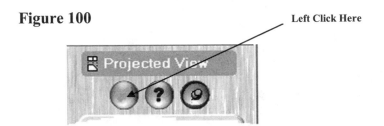

112. If the views were inadvertently placed too close together, move the cursor to any of the views causing red dots to appear as shown in Figure 101.

Figure 101

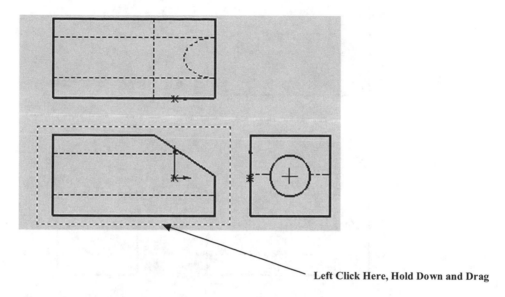

Left Click Here, Hold Down and Drag

113. Move the cursor to the red dots and left click (holding the left mouse button down). Drag the cursor to move the view to the desired location.

114. Move the cursor to the upper left portion of the screen and left click on the green checkmark as shown in Figure 102.

Figure 102

Left Click Here

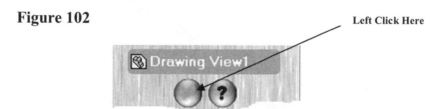

173

115. Your screen should look similar to Figure 103.

Figure 103

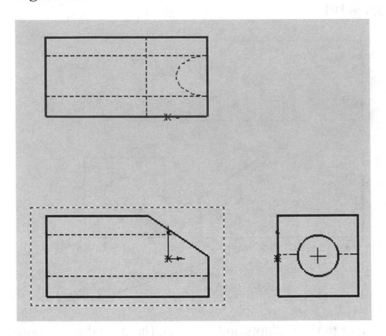

116. Save the part file as Part1.SLDDRW for easy retrieval. This part will be used in the following chapter.

Drawing Activities

Use these problems from Chapters 1 and 2 to create 3 view orthographic view detail drawings.

Problem 1

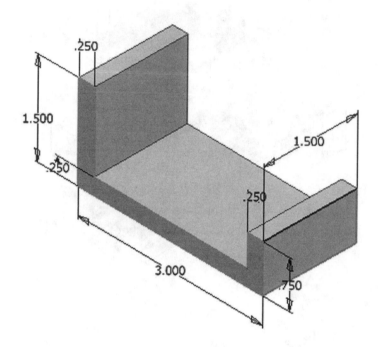

Problem 2

Extrude Center Section .25 Deep

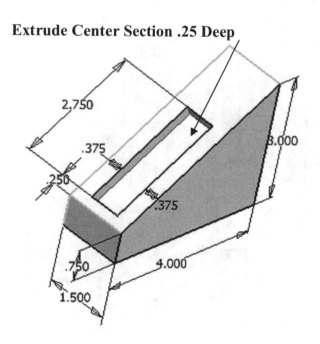

Problem 3

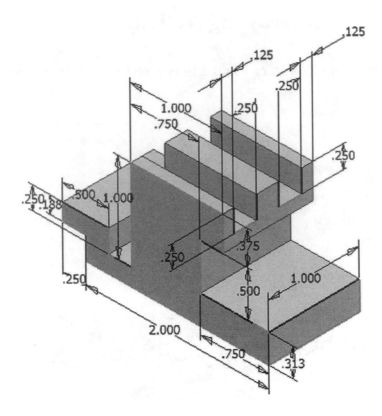

Problem 4

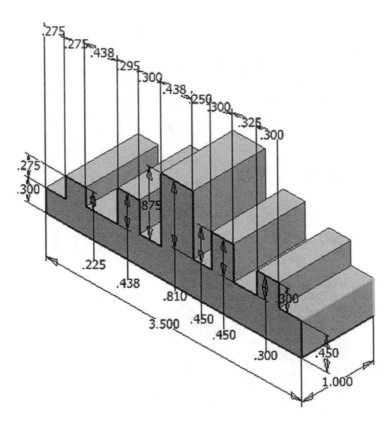

Problem 5

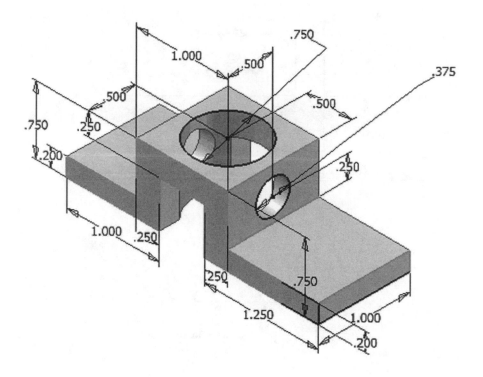

Problem 6

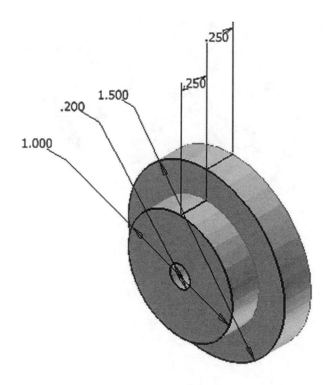

Problem 7

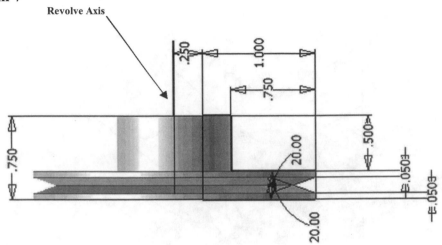

Problem 8

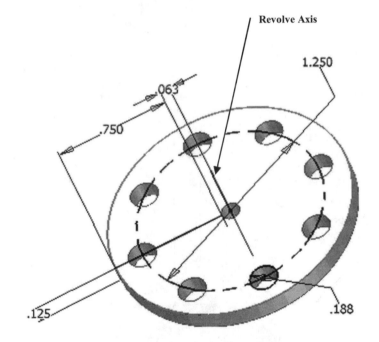

Chapter 4 Advanced Detail Drawing Procedures

Objectives:

- Create an Auxiliary View using the Auxiliary View command
- Dimension views using the Smart Dimension command
- Create a Section View using the Section View command
- Create Text using the Note command

Chapter 4 includes instruction on how to create the drawings shown below.

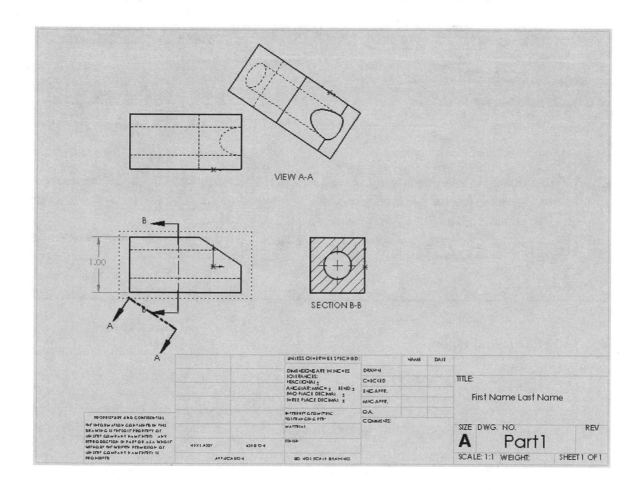

1.   Start SolidWorks by referring to "Chapter 1 Getting Started".

2.   After SolidWorks is running, open the .SLDDRW file that was created in Chapter 3. Move the cursor to the upper left corner of the screen and left click on "Open" icon as shown in Figure 1.

Figure 1   Left Click Here

3.   The Open dialog box will appear. Left click on the drawing that was created in Chapter 3 as shown in Figure 2.

Figure 2   Left Click Here

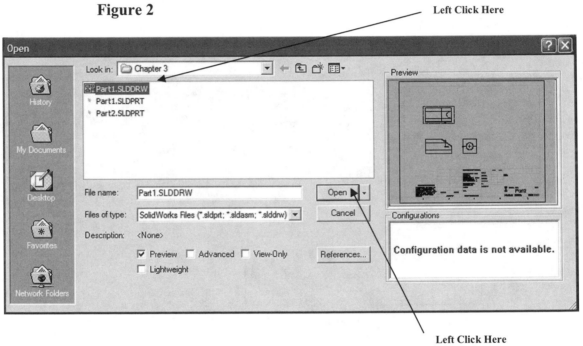

Left Click Here

4.   Left Click on **Open**.

5.   After the .SLDDRW file is open, move the views closer to each other to provide additional room on the drawing. Start by moving the cursor over the top view. Dots will appear around the view. Left click (holding the left mouse button down) on the dots and drag the view down closer to the front view as shown in Figure 3.

Figure 3   Left Click Here, Hold Down and Drag

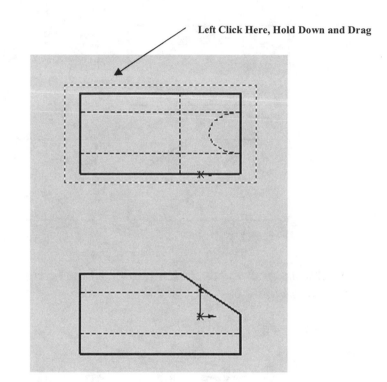

6.   Move the side view closer to the front view.   Start by moving the cursor over the side view. Dots will appear around the view. Left click (holding the left mouse button down) on the dots and drag the view closer to the front view as shown in Figure 4.

Figure 4   Left Click Here, Hold Down and Drag

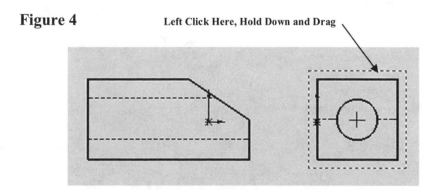

7.   This will provide more room to work. Your screen should look similar to Figure 5.

Figure 5

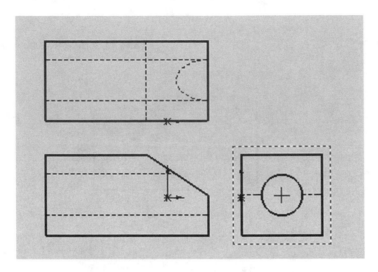

8.   Move the cursor to the upper left portion of the screen and left click on **Drawings** as shown in Figure 6.

Figure 6   Left Click Here

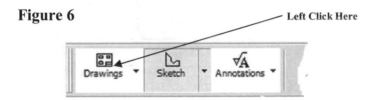

9.   Notice that the commands at the top of the screen are different as shown in Figure 7.

Figure 7

10.   Left click on **Auxiliary View** as shown in Figure 8.

Figure 8   Left Click Here

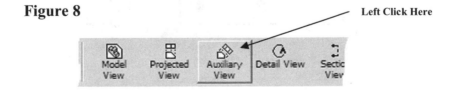

11. Move the cursor to the wedge line and left click as shown in Figure 9.

Figure 9

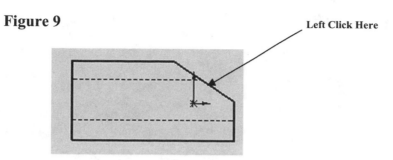

12. SolidWorks will create an auxiliary view from the selected surface. The view will be attached to the cursor as shown in Figure 10.

Figure 10

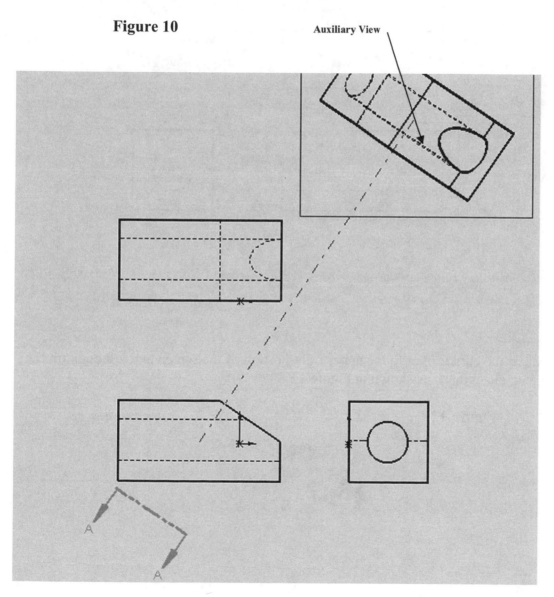

13. Drag the cursor towards the other views and left click as shown in Figure 11.

Figure 11

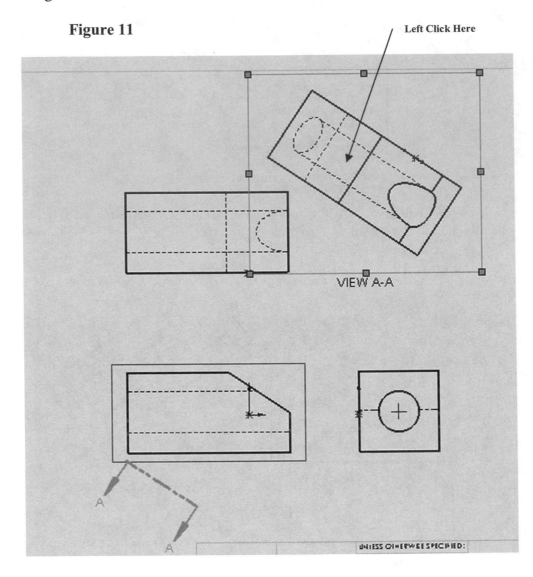

14. Move the cursor over to the upper right portion of the screen and left click on the green checkmark as shown in Figure 12.

Figure 12

15. Your screen should look similar to Figure 13.

Figure 13

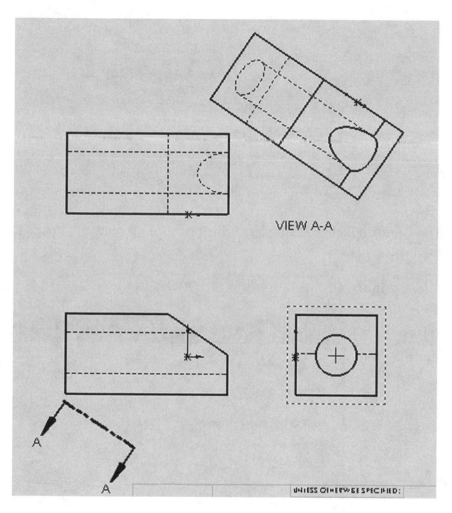

16. Move the cursor to the side view. Red dots will appear as shown in Figure 14.

Figure 14

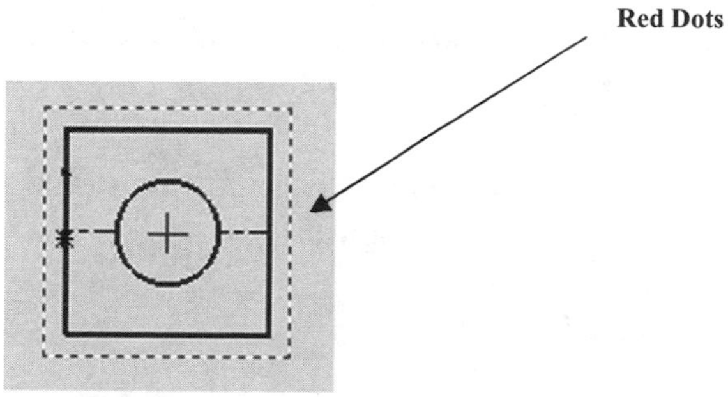

185

17. Right click on the view. A pop up menu will appear. Left click on **Delete** as shown in Figure 15.

Figure 15

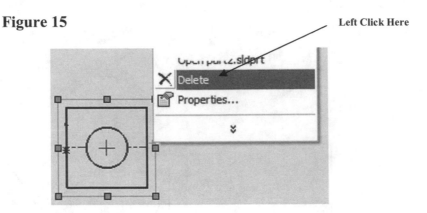

Left Click Here

18. The Confirm Delete dialog box will appear. Left click on **Yes** as shown in Figure 16.

Figure 16

Left Click Here

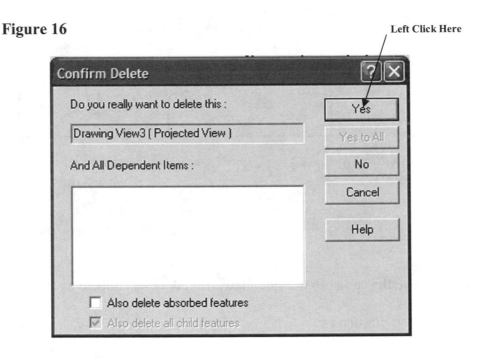

19. Move the cursor to the upper middle portion of the screen and left click on **Section View** as shown in Figure 17.

Figure 17

Left Click Here

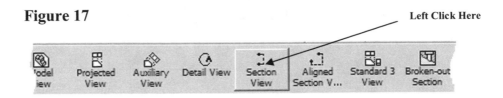

186

20. Move the cursor over the front view. Red dots will appear around the view. Left click above the top of the part as shown in Figure 18.

Figure 18

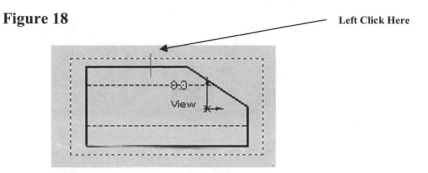

Left Click Here

21. Drag the cursor down across the part and left click as shown in Figure 19.

Figure 19

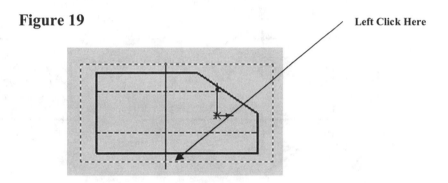

Left Click Here

22. Drag the cursor to the right. The section view will be attached to the cursor. Place the section view to the side of the front view and left click as shown in Figure 20.

Figure 20

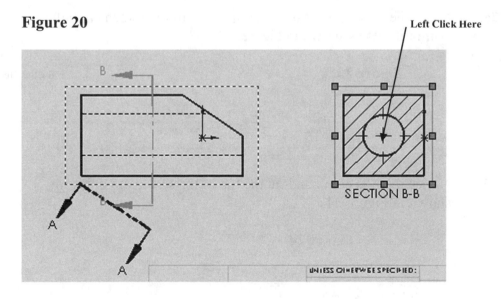

Left Click Here

SECTION B-B

23. Move the cursor to the upper left portion of the screen and left click on the green checkmark as shown in Figure 21.

Figure 21

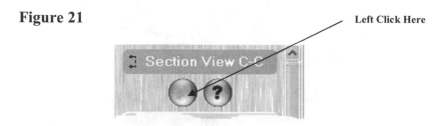

24. SolidWorks will create a section view to the right as shown in Figure 22.

Figure 22

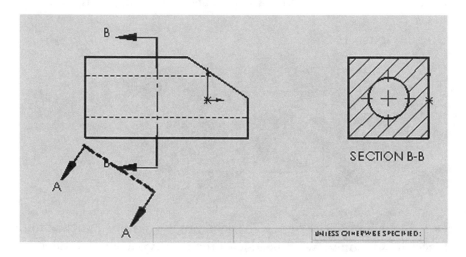

25. Move the cursor to the upper left portion of the screen and left click on **Annotations** as shown in Figure 23.

Figure 23

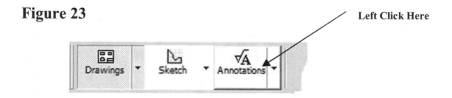

26. Notice that the commands at the top of the screen are different as shown in Figure 24.

Figure 24

27. Annotations commands are typically where "drafting" activities are performed.

28. Move the cursor to the upper left portion of the screen and left click on **Smart Dimension** as shown in Figure 25.

Figure 25

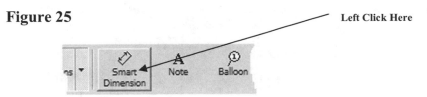

29. Move the cursor over the left side vertical line until it turns red as shown in Figure 26. Select the line by left clicking anywhere on the line **or** on each of the end points. To use the end points of the line, move the cursor over one of the end points. A small red dot will appear. Left click once and move the cursor to the other end point. After the red dot appears, left click once. The dimension will be attached to the cursor.

Figure 26

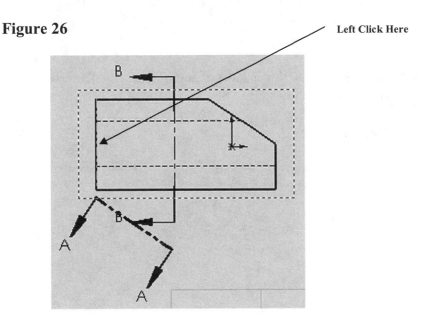

30. Move the cursor to where the dimension will be placed and left click once as shown in Figure 27.

Figure 27

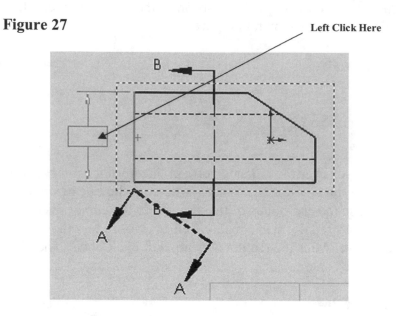

31. Move the cursor to the upper left portion of the screen and left click on the green checkmark as shown in Figure 28.

Figure 28

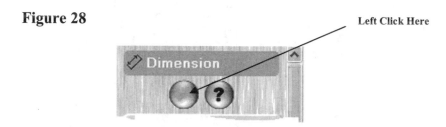

32.   Your screen should look similar to Figure 29.

Figure 29

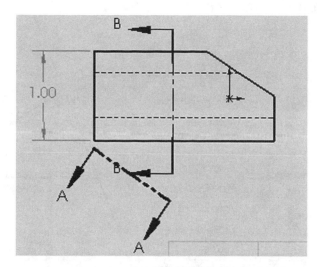

33.   Finish dimensioning the part to your own satisfaction. When the part is satisfactorily dimensioned, save the file to a location where it can be easily retrieved.

34.   To hide any dimension, move the cursor over the dimension. A red box will appear around the dimension as shown in Figure 30.

Figure 30

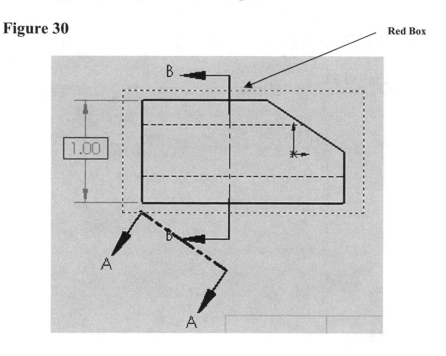

35. Right click on the dimension. A pop up menu will appear. Left click on **Hide** as shown in Figure 31.

Figure 31

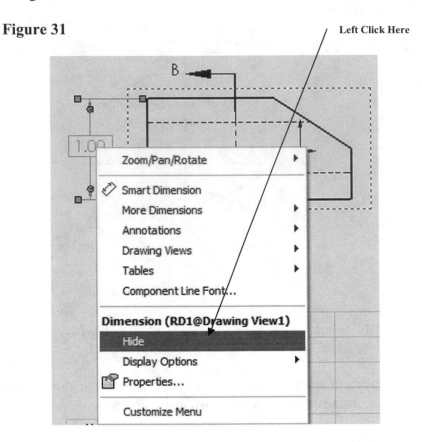

36. Left click on **Note** as shown in Figure 32.

Figure 32

37. Move the cursor to the title block location. A small box will be attached to the cursor. Left click as shown in Figure 33.

Figure 33

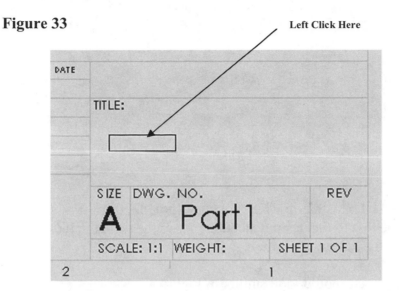

38. The cursor will begin blinking. Enter your first name and last name. After the text has been entered, move the cursor to the upper left portion of the screen and left click on the green checkmark as shown in Figure 34.

Figure 34

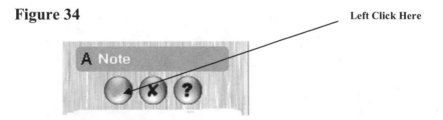

39. Text will appear in the title block as shown in Figure 35.

Figure 35

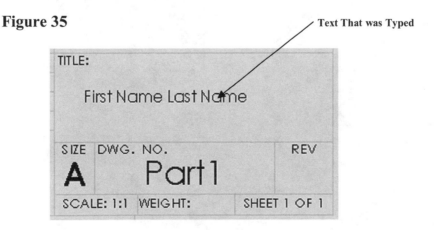

40. If the text needs to be moved, move the cursor over the text causing a red box to appear as shown in Figure 36.

Figure 36

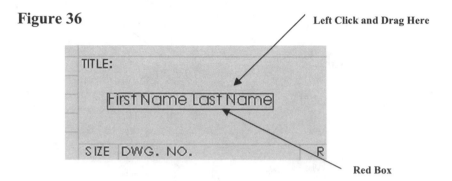

41. While the text is highlighted, left click (holding the left mouse button down) and drag the text to the desired location. After the text is in the desired location, release the left mouse button.

42. Your screen should look similar to Figure 37.

Figure 37

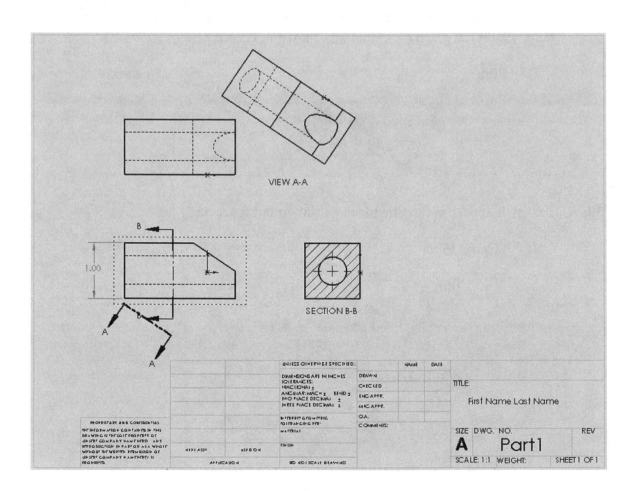

43.   Save the current sheet where it can be easily retrieved.

Drawing Activities

Create Section View Drawings for the following:

Problem 1

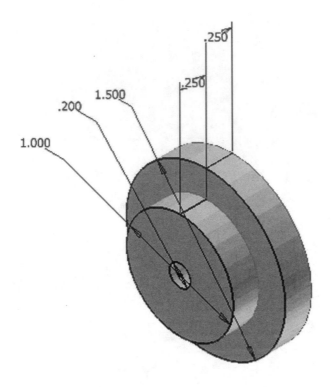

Problem 2

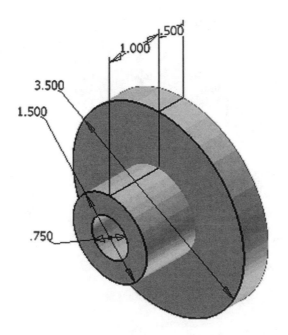

Create Auxiliary View Drawings for the following:

Problem 3

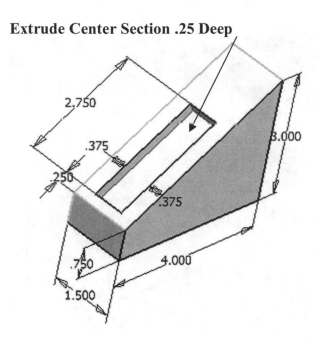

Extrude Center Section .25 Deep

Problem 4

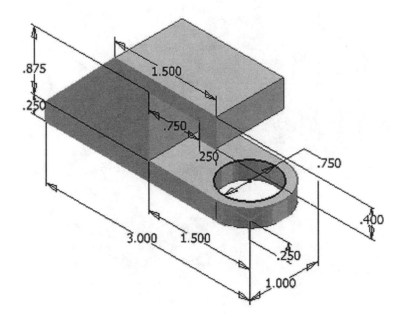

197

Create Section View Drawings for the following:

Problem 5

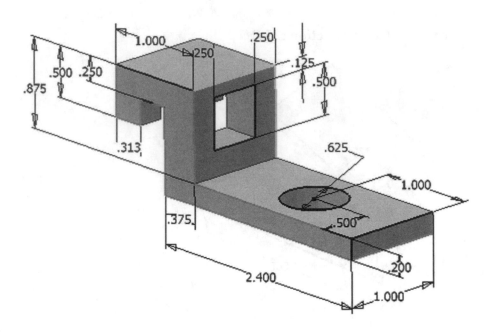

Problem 6

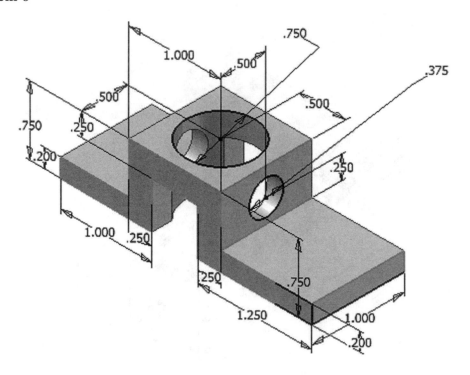

Create Section View Drawings for the following:

Problem 7

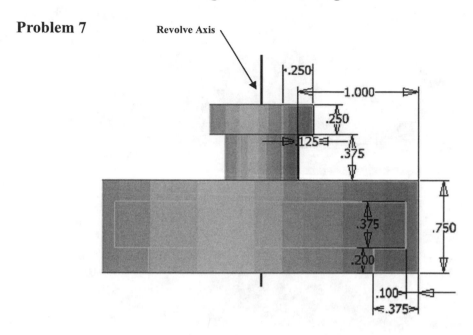

Problem 8

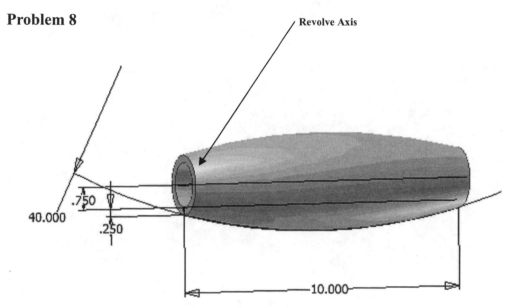

Chapter 5 Learning To Edit Existing Solid Models

Objectives:

- Design a simple part
- Learn to use the Circular Pattern command
- Learn to edit a part using the Circular Pattern command
- Edit the part using the Sketch commands
- Edit the part using the Extrude command
- Edit the part using the Fillet command

Chapter 5 includes instruction on how to design and edit the part shown below.

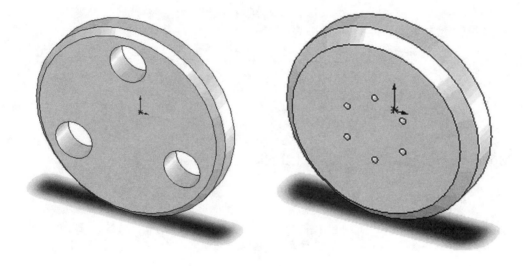

1.   Start SolidWorks by referring to "Chapter 1 Getting Started".

2.   After SolidWorks is running, begin a new sketch.

3.   Move the cursor to the upper left corner of the screen and left click on **Circle** as shown in Figure 1.

Figure 1   Left Click Here

4.   Move the cursor to the center of the screen and left click once. This will be the center of the circle as shown in Figure 2.

Figure 2   Left Click Here

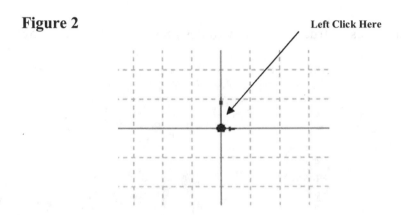

5.   Move the cursor to the right and left click once as shown in Figure 3.

Figure 3   Left Click Here

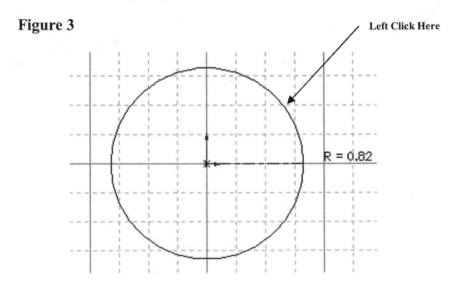

R = 0.82

6. Right click anywhere on the drawing. A pop up menu will appear. Left click on **Select** as shown in Figure 4.

Figure 4

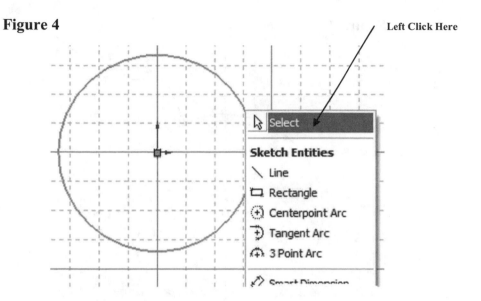

Left Click Here

7. Move the cursor to the middle left portion of the screen and left click on **Smart Dimension** as shown in Figure 5.

Figure 5

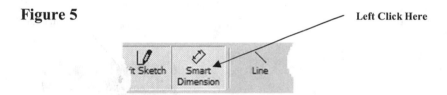

Left Click Here

8. Move the cursor over the edge (not center) of the circle until it turns red. Left click once as shown in Figure 6. The dimension will be attached to the cursor.

Figure 6

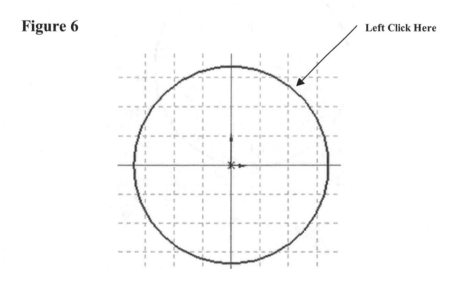

Left Click Here

9. Move the cursor to where the dimension will be placed and left click once as shown in Figure 7.

Figure 7

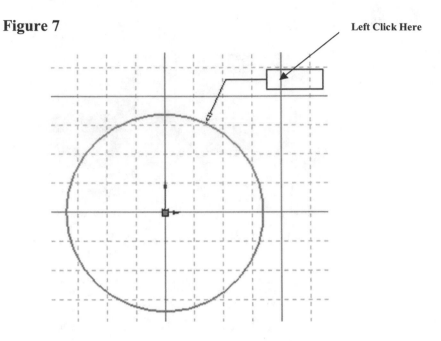

10. The Modify dialog box will appear as shown in Figure 8.

Figure 8

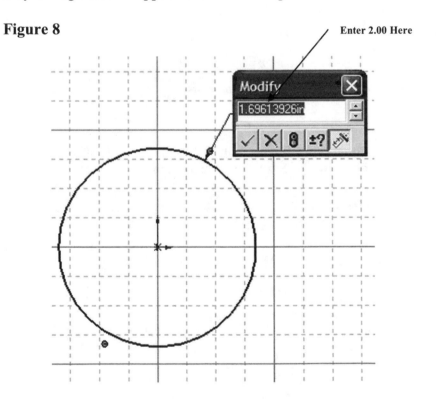

11. To edit the dimension, type **2.00** in the Modify dialog box (while the current dimension is highlighted) and left click on the green checkmark as shown in Figure 9.

Figure 9

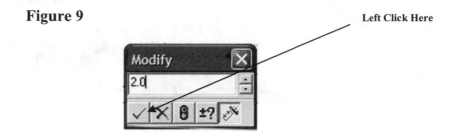

Left Click Here

12. The dimension of the circle will become 2.00 inches as shown in Figure 10.

Figure 10

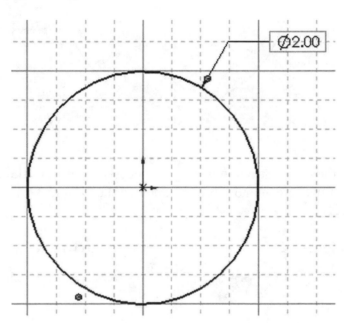

13.   Right click anywhere around the drawing. A pop up menu will appear. Left click on **Select** as shown in Figure 11.

Figure 11

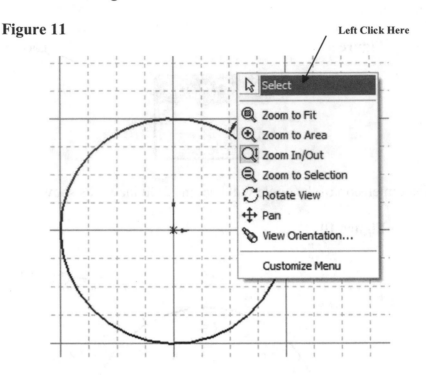

14.   To view the entire drawing, it may be necessary to move the cursor to the middle portion of the screen and left click once on the "Zoom To Fit" icon as shown in Figure 12.

Figure 12   Left Click Here

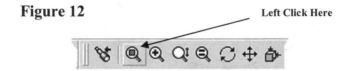

15.   The drawing will "fill up" the entire screen. If the drawing is still too large, left click on the "Zoom In/Out" icon as shown in Figure 13. After selecting the "Zoom In/Out" icon, hold the left mouse button down and drag the cursor up and down to achieve the desired view of the sketch.

Figure 13   Left Click Here

16. After the sketch is complete it is time to extrude the sketch into a solid. Right click anywhere on the drawing. A pop up menu will appear. Left click on **Select** as shown in Figure 14.

Figure 14

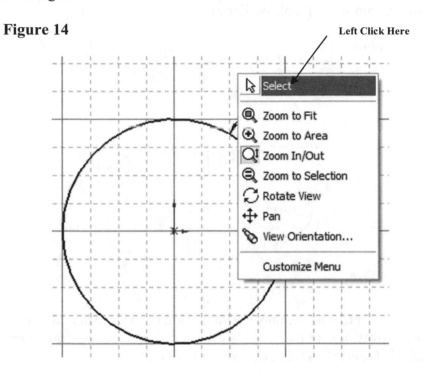

17. After you have verified that no commands are active, right click anywhere on the sketch. A pop up menu will appear. Left click on **Exit Sketch** as shown in Figure 15.

Figure 15

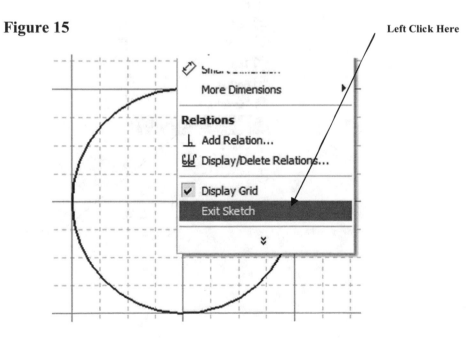

18. SolidWorks is now out of the Sketch commands and into the Features commands. Notice that the commands at the top of the screen are now different. To work in the Features commands a sketch must be present and have no opens (non-connected lines). If there are any opens in the sketch an error message will appear. Your screen should look similar to Figure 16.

Figure 16

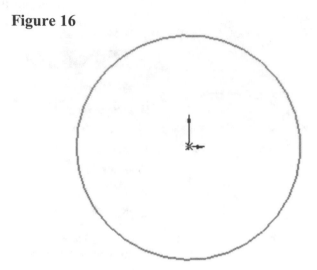

19. Move the cursor to the upper right portion of the screen and left click on the drop down arrow next to the "Standard Views" icon. A drop down menu will appear. Left click on **Trimetric** as shown in Figure 17.

Figure 17

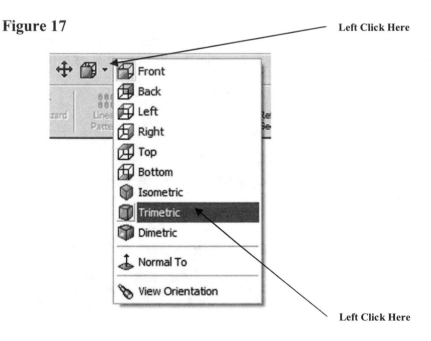

20. The view will become trimetric as shown in Figure 18.

Figure 18

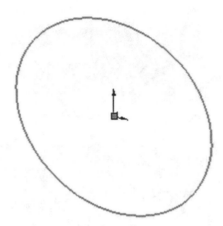

21. Move the cursor to the upper left portion of the screen and left click on **Extruded Boss/Base** as shown in Figure 19.

Figure 19

Left Click Here

22. A preview of the extrusion will be displayed. If SolidWorks gave you an error message, there are opens (non-connected lines) somewhere on the sketch. Check each intersection for opens by using the **Extend** and **Trim** commands.

23. Move the cursor to the middle left portion of the screen and enter **.25** next to D1 as shown in Figure 20.

Figure 20

Enter .25 Here

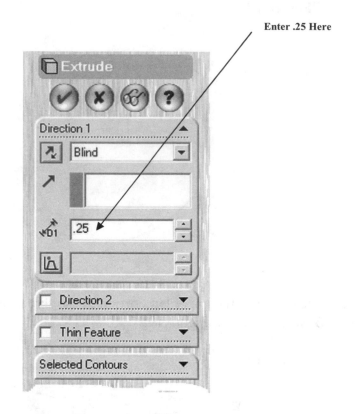

24. Move the cursor to the upper left portion of the screen and left click on the green checkmark as shown in Figure 21.

Figure 21

Left Click Here

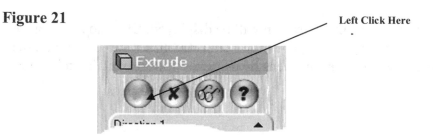

25. Your screen should look similar to Figure 22.

Figure 22

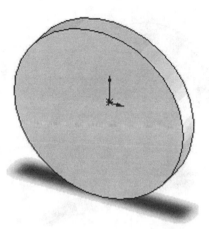

26. Move the cursor to the front surface causing the edges of the surface to turn red. Right click on the surface once. A pop up menu will appear. Left click on **Insert Sketch** as shown in Figure 23.

Figure 23   Left Click Here

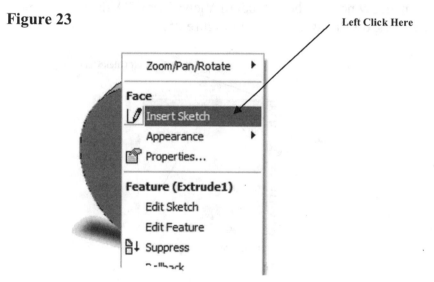

27. SolidWorks will start a new sketch on the selected surface as shown in Figure 24.

Figure 24

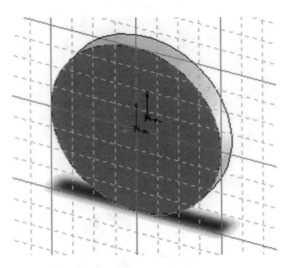

28. Move the cursor to the upper right portion of the screen and left click on the drop down arrow next to the "Standard Views" icon. A drop down menu will appear. Left click on **Front** as shown in Figure 25.

Figure 25

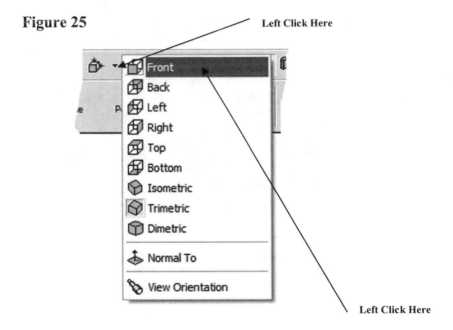

29.   The part will rotate providing a perpendicular view of the surface as shown in Figure 26.

Figure 26

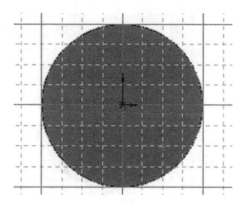

30.   Move the cursor to the upper left corner of the screen and left click on **Circle** as shown in Figure 27.

Figure 27   Left Click Here

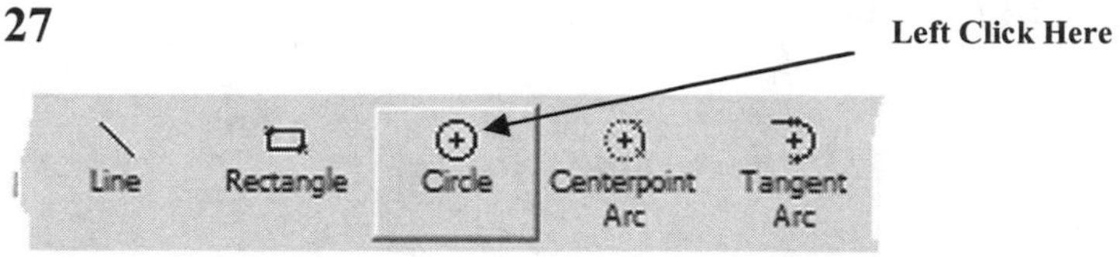

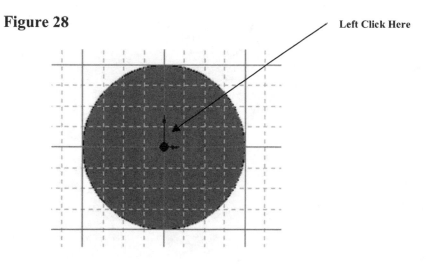

31.   Move the cursor to the center of the circle. A red dot will appear as shown in Figure 28.

Figure 28   Left Click Here

32. After the red dot appears, move the cursor straight up causing a blue dashed line to appear. Left click once as shown in Figure 29.

Figure 29

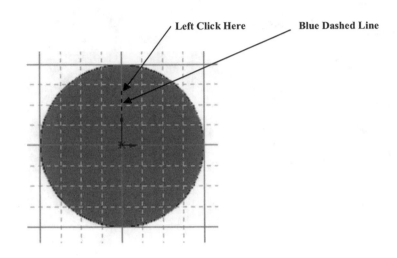

33. Move the cursor to the side. A circle will form. Left click once as shown in Figure 30.

Figure 30

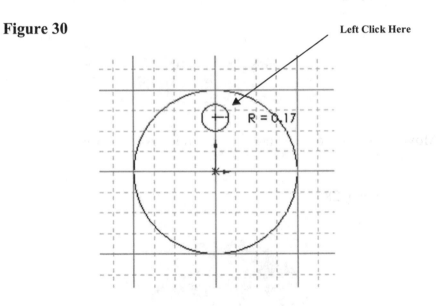

34. Right click anywhere on the drawing. A pop up menu will appear. Left click on **Select** as shown in Figure 31.

Figure 31

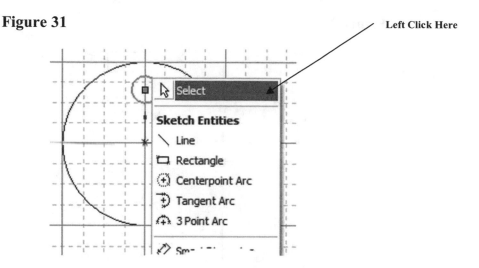

35. Move the cursor to the middle left portion of the screen and left click on **Smart Dimension** as shown in Figure 32.

Figure 32

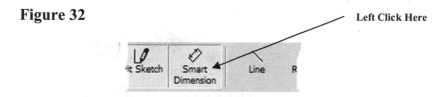

36. Move the cursor over the edge (not center) of the circle until it turns red as shown in Figure 33. Select the circle by left clicking anywhere on the edge. The dimension will be attached to the cursor.

Figure 33

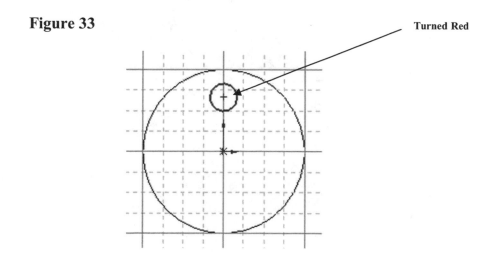

37. Move the cursor to where the dimension will be placed and left click once as shown in Figure 34.

Figure 34

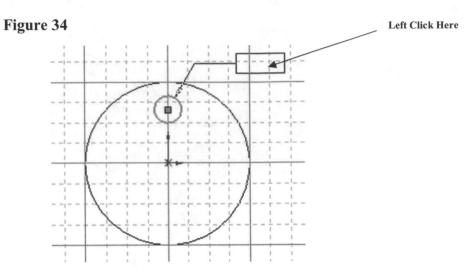

Left Click Here

38. The Modify dialog box will appear as shown in Figure 35.

Figure 35

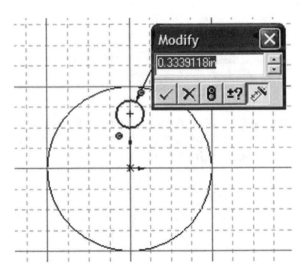

39. To edit the dimension, type **.375** in the Modify dialog box (while the current dimension is highlighted) and left click on the green checkmark as shown in Figure 36.

Figure 36

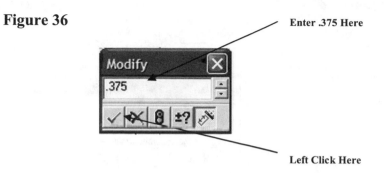

Enter .375 Here

Left Click Here

40. The dimension of the circle will become .375 inches as shown in Figure 37. Use the Zoom icons to zoom out if necessary.

Figure 37

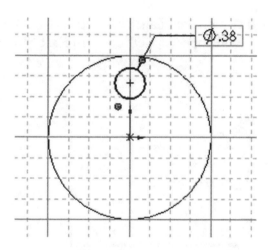

41.   Right click anywhere around the drawing. A pop up menu will appear. Left click
on **Select** as shown in Figure 38.

Figure 38

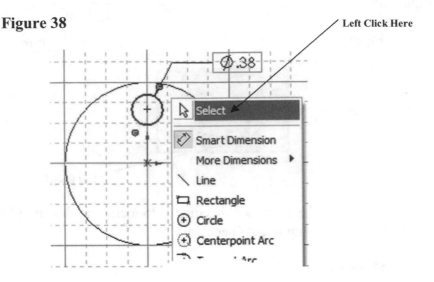

42.   After you have verified that no commands are active, right click anywhere on the
sketch. A pop up menu will appear.   Left click on **Exit Sketch** as shown in
Figure 39.

Figure 39

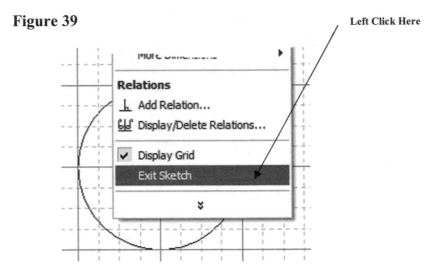

43. SolidWorks is now out of the Sketch commands and into the Features commands. Notice that the commands at the top of the screen are now different. To work in the Features commands a sketch must be present and have no opens (non-connected lines). If there are any opens in the sketch an error message will appear. Your screen should look similar to Figure 40.

Figure 40

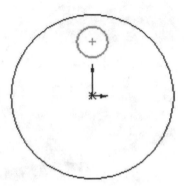

44. Move the cursor to the upper right portion of the screen and left click on the drop down arrow next to the "Standard Views" icon. A drop down menu will appear. Left click on **Trimetric** as shown in Figure 41.

Figure 41

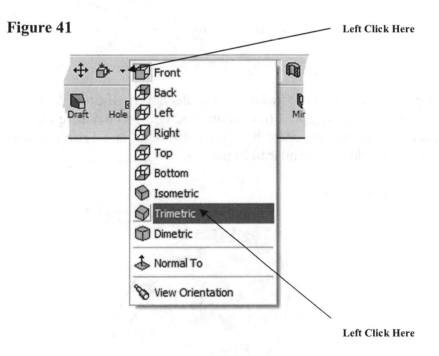

Left Click Here

Left Click Here

45. The view will become trimetric as shown in Figure 42.

Figure 42

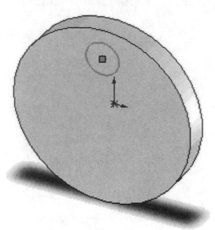

46. Move the cursor to the upper left portion of the screen and left click on **Extruded Cut** as shown in Figure 43.

Figure 43   **Left Click Here**

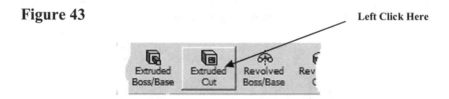

47. A preview of the cut extrusion will be displayed. If SolidWorks gave you an error message, there are opens (non-connected lines) somewhere on the sketch. Check each intersection for opens by using the **Extend** and **Trim** commands. Your screen should look similar to Figure 44.

Figure 44

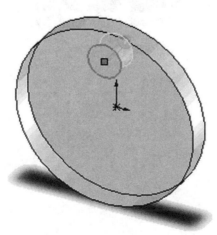

48. Enter **.25** next to D1 as shown in Figure 45.

Figure 45

Enter .25 Here

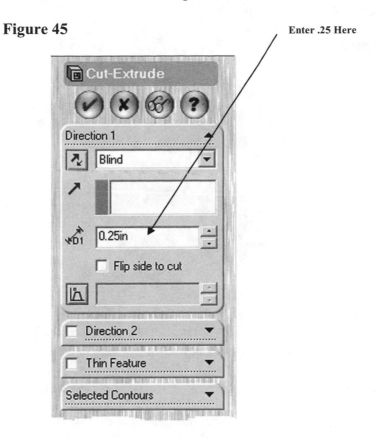

49. Move the cursor to the upper left portion of the screen and left click on the green checkmark as shown in Figure 46.

Figure 46

Left Click Here

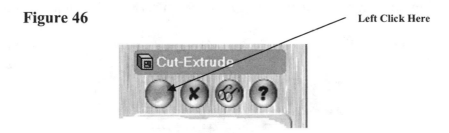

50. Your screen should look similar to Figure 47. You may have to use the zoom out command to view the entire part.

Figure 47

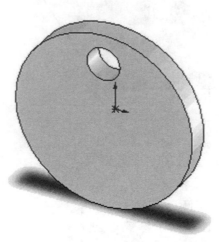

51. Move the cursor to the upper right portion of the screen and left click on **Reference Geometry** as shown in Figure 48.

Figure 48

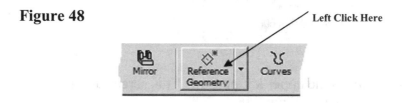

52. A drop down menu will appear. Left click on **Axis** as shown in Figure 49.

Figure 49

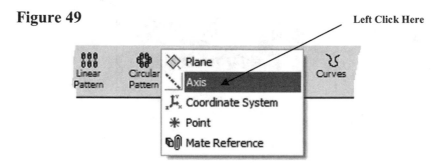

53. Move the cursor to the side of the part causing both edges to turn red. Left click once as shown in Figure 50.

Figure 50

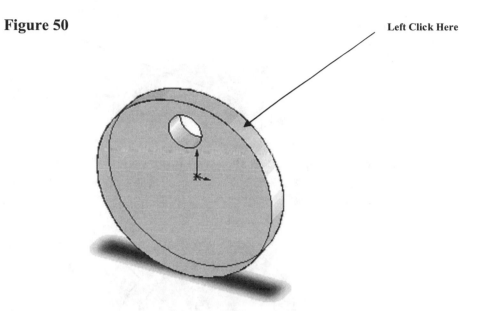

Left Click Here

54. SolidWorks will project a center axis as shown in Figure 51. This axis will be used when creating a circular pattern.

Figure 51

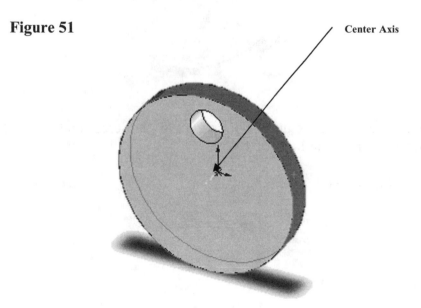

Center Axis

55.   Move the cursor to the middle left portion of the screen and left click on the green checkmark as shown in Figure 52.

Figure 52

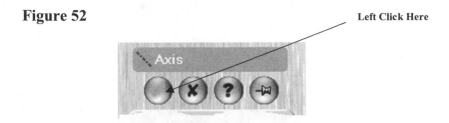

Left Click Here

56.   The axis will be displayed as shown in Figure 53.

Figure 53

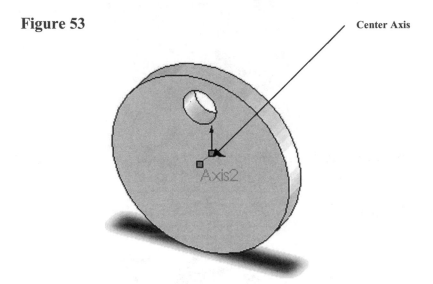

Center Axis

57.   Move the cursor to the upper right portion of the screen and left click on **Circular Pattern** as shown in Figure 54.

Figure 54

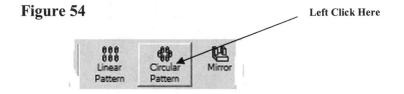

Left Click Here

58. Move the cursor to the left portion of the screen and enter **3** for the number of holes and left click in the box next to the text "Equal spacing" as shown in Figure 55.

Figure 55

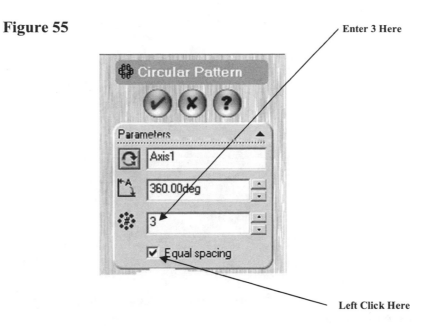

Enter 3 Here

Left Click Here

59. Move the cursor to the center of the circle (hole) as shown in Figure 56. The edges will become red dashed lines. SolidWorks will only find the circle (hole) if the view is trimetric or isometric.

Figure 56

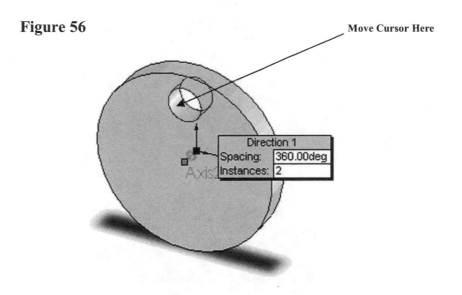

Move Cursor Here

60. Left click inside the circle (hole). The inside of the hole will turn green. A preview of all holes will be displayed as shown in Figure 57.

Figure 57

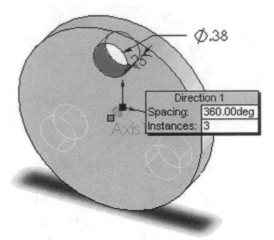

61. Move the cursor to the left portion of the screen and left click on the green checkmark as shown in Figure 58.

Figure 58   Left Click Here

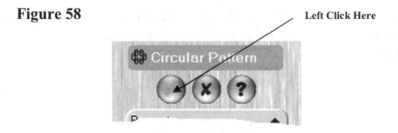

62. Your screen should look similar to Figure 59.

Figure 59

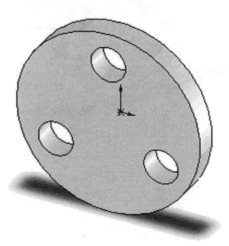

63. Move the cursor to the upper middle portion of the screen and left click on **Chamfer** as shown in Figure 60.

Figure 60

64. Enter **.0625** for the Radius as shown in Figure 61. Left click next to Full Preview. Press **Enter** on the keyboard.

Figure 61

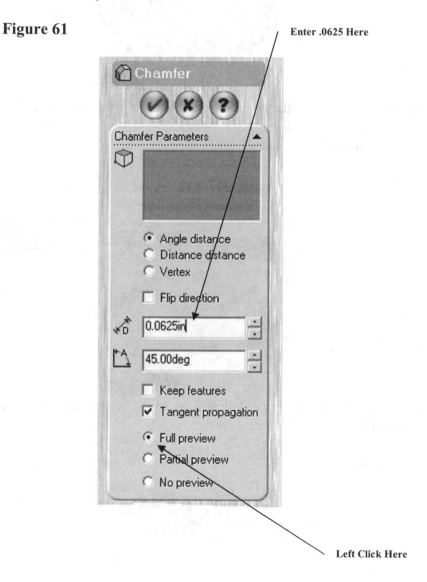

65.   Move the cursor over the edge of the part causing it to turn red as shown in Figure 62. Left click once.

Figure 62

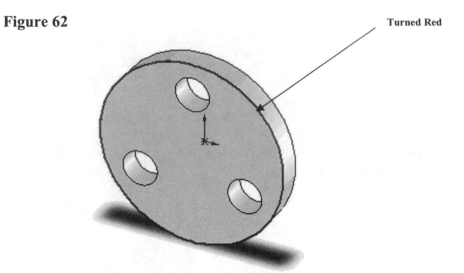

66.   A preview of the chamfer will be displayed as shown in Figure 63.

Figure 63

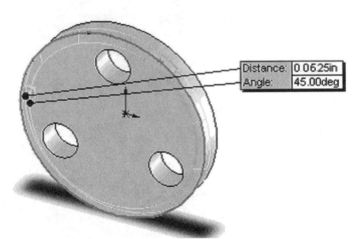

67.   Move the cursor to the upper left portion of the screen and left click on the green checkmark as shown in Figure 64.

Figure 64

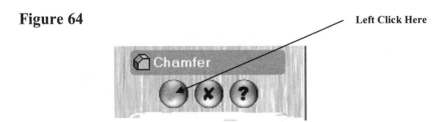

68. Your screen should look similar to Figure 65.

Figure 65

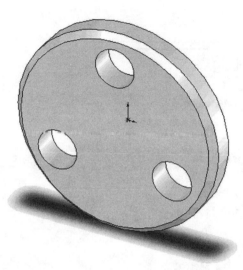

69. If for some reason a change needs to be made to this part, it can be accomplished by editing either a sketch or a feature located in the Part Tree at the lower left corner of the screen as shown in Figure 66.

Figure 66   **Part Tree Location**

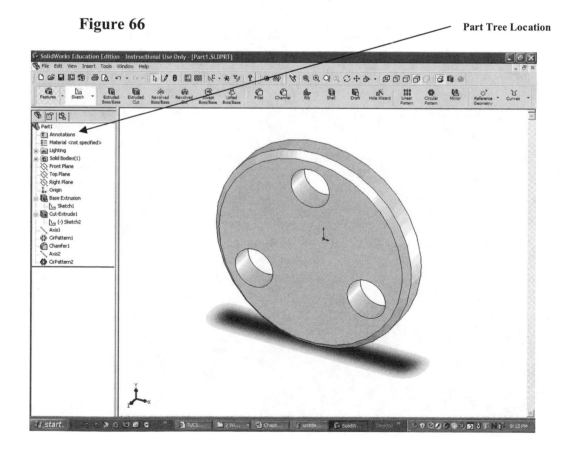

70.   The Part Tree is shown in Figure 67. Left click on each of the "plus" signs in the part tree. The tree will expand showing more details for part construction.

Figure 67

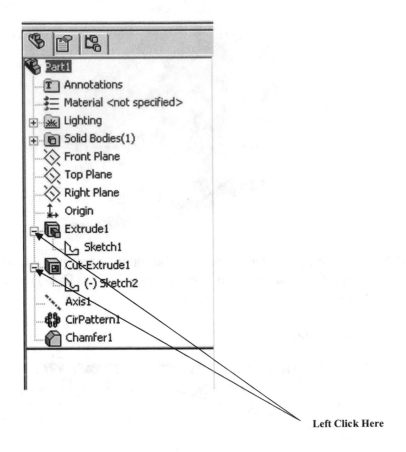

Left Click Here

71. Move the cursor over the text "Sketch1". The text will become highlighted as shown in Figure 68.

Figure 68

Highlighted Text

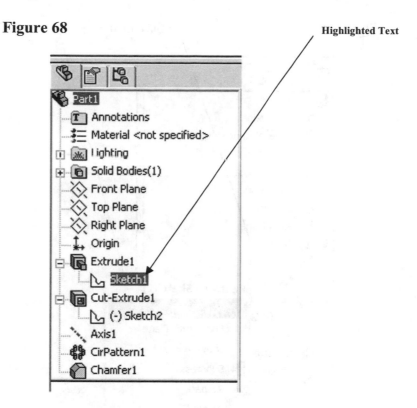

72. The original sketch will also appear as shown in Figure 69.

Figure 69

Original Sketch

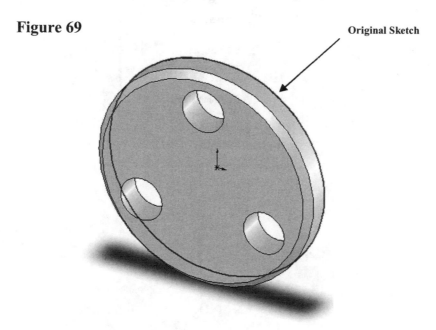

73. Right click on **Sketch1**. A pop up menu will appear. Left click on **Edit Sketch** as shown in Figure 70.

Figure 70

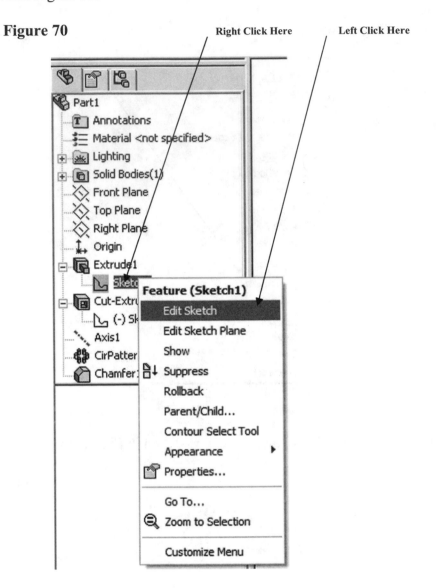

74. The original sketch will appear as shown in Figure 71.

Figure 71

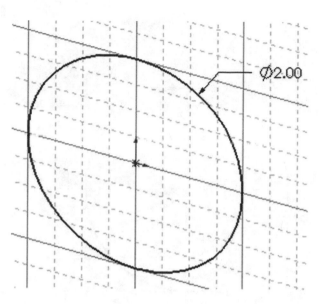

75. Move the cursor to the upper middle portion of the screen and left click on the drop down arrow next to the "Standard Views" icon. A drop down menu will appear. Left click on **Front** as shown in Figure 72.

Figure 72

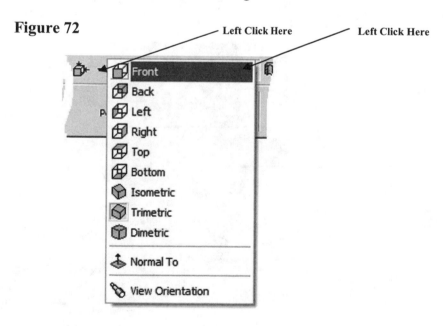

76. A perpendicular view of the sketch will be displayed. The sketch will look similar to when the sketch was first constructed. Your screen should look similar to Figure 73.

Figure 73

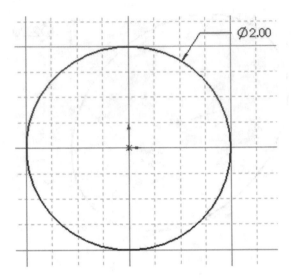

77. Start by modifying the diameter of the part. Double click on the overall dimension. The Modify dialog box will appear as shown in Figure 74.

Figure 74

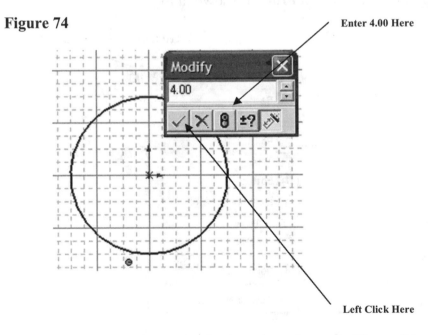

78. Enter **4.00** and left click on the green checkmark as shown in Figure 74.

234

79. The diameter of the part will increase to 4.00 as shown in Figure 75.

Figure 75

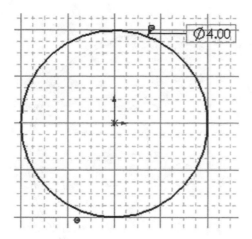

80. Move the cursor to the upper left portion of the screen and left click on the green checkmark as shown in Figure 76.

Figure 76   Left Click Here

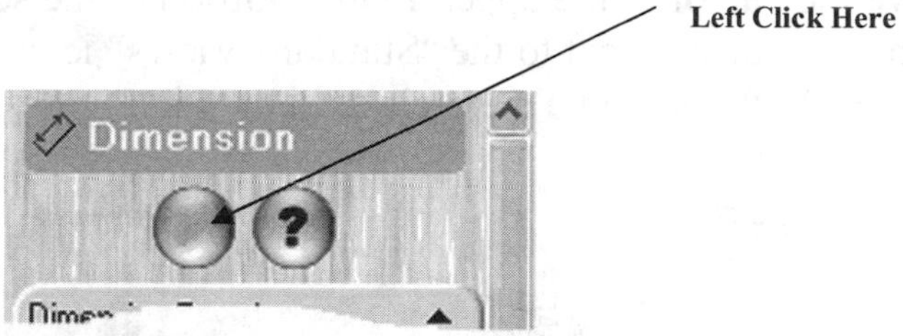

81. Right click anywhere on the screen. A pop up menu will appear. Left click on **Exit Sketch** as shown in Figure 77.

Figure 77   Left Click Here

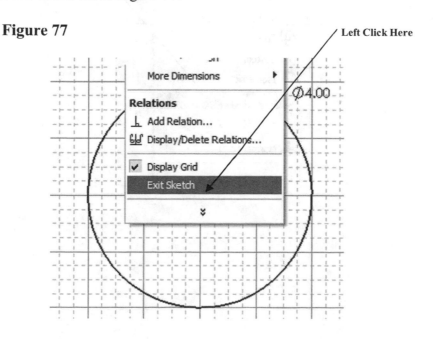

235

82.   SolidWorks will automatically update the part to reflect the changes made in the sketch. It will not be necessary to extrude the part again. Your screen should look similar to Figure 78.

Figure 78

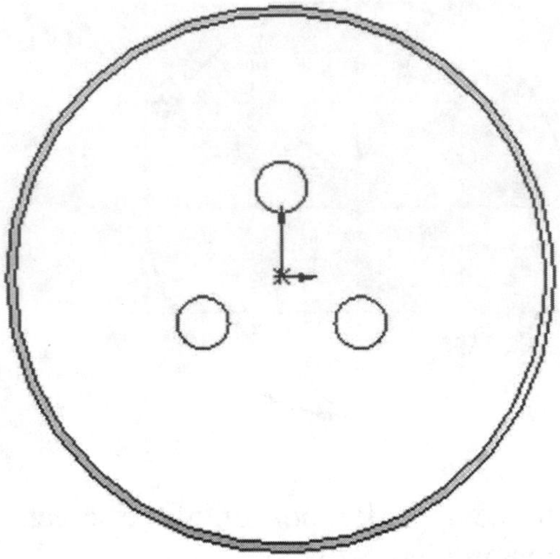

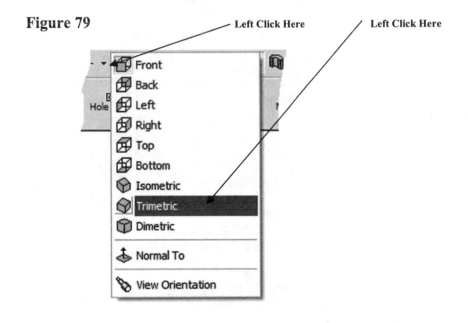

83.   Move the cursor to the upper middle portion of the screen and left click on the drop down arrow next to the "Standard Views" icon. A drop down menu will appear. Left click on **Trimetric** as shown in Figure 79.

Figure 79

84. Your screen should look similar to Figure 80.

Figure 80

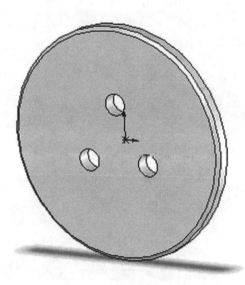

85. Move the cursor to the part tree. Right click once on **Extrude1**. A pop up menu will appear. Left click on **Edit Feature** as shown in as shown in Figure 81.

Figure 81

Left Click Here

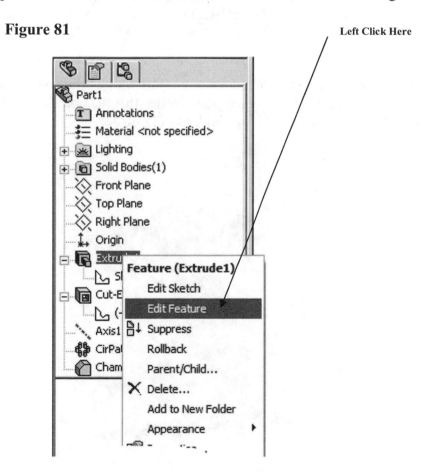

86. Move the cursor to the middle left portion of the screen. Thc Extrude1 dialog box will appear. Enter **.750** for the extrusion distance and left click on the green checkmark as shown in Figure 82.

Figure 82

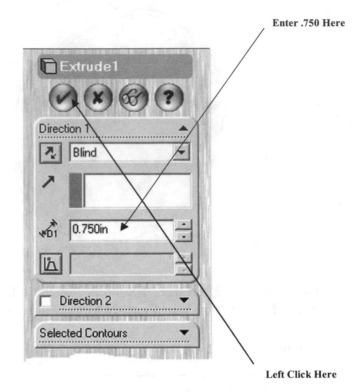

Enter .750 Here

Left Click Here

87. SolidWorks will automatically update the part without the need to repeat any of the steps that created the original part. Notice that the holes are no longer "thru" holes. Your screen should look similar to Figure 83.

Figure 83

No Longer Thru Holes

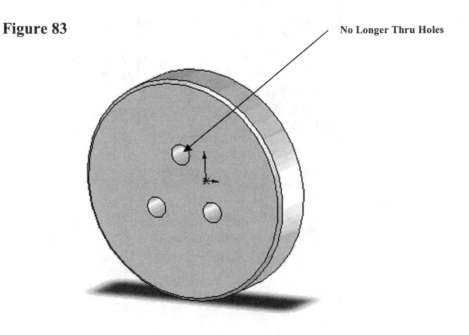

88.　Move the cursor to the upper left portion of the screen where the part tree is located. Right click once on **Sketch2**. A pop up menu will appear. Left click on **Edit Sketch** as shown in Figure 84.

Figure 84

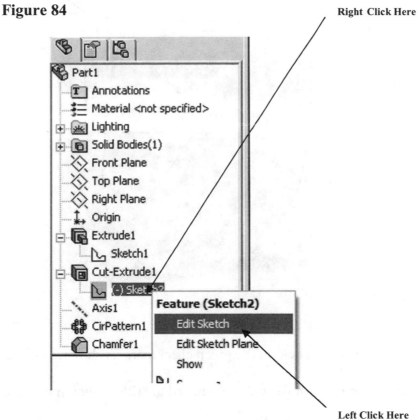

89.　The original sketch will appear as shown in Figure 85.

Figure 85

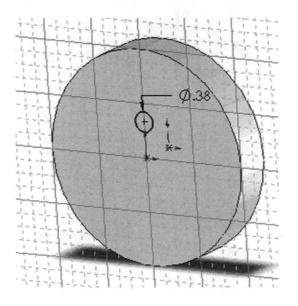

90. Modify the diameter of the holes. Double click on the overall dimension. The Modify dialog box will appear as shown in Figure 86.

Figure 86

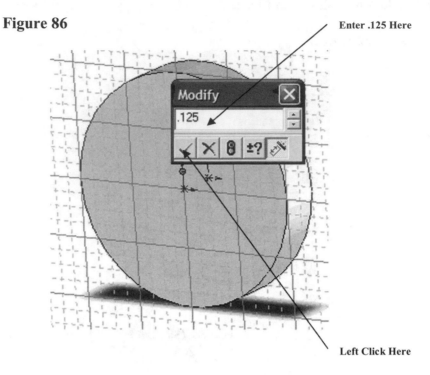

Enter .125 Here

Left Click Here

91. Enter **.125** and left click on the green checkmark as shown in Figure 86.

92. The diameter of the holes will be reduced to .125 as shown in Figure 87.

Figure 87

Hole Diameter Reduced to .125

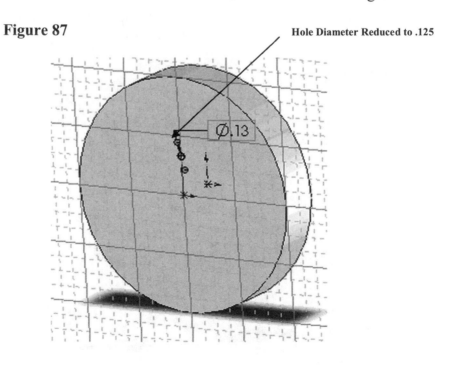

93.   Move the cursor to the upper left portion of the screen and left click on the green checkmark as shown in Figure 88.

Figure 88

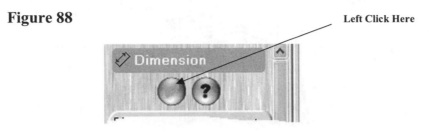

Left Click Here

94.   Right click anywhere around the drawing. A pop up menu will appear. Left click on **Exit Sketch** as shown in Figure 89.

Figure 89

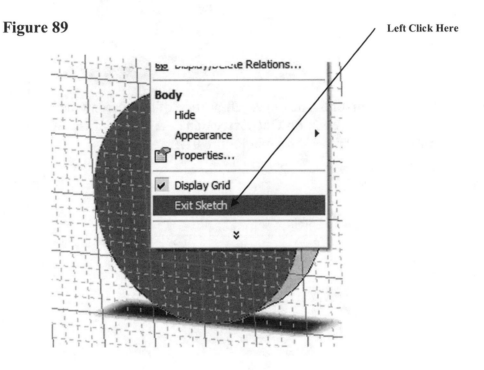

Left Click Here

95. SolidWorks will automatically update the part as shown in Figure 90.

Figure 90

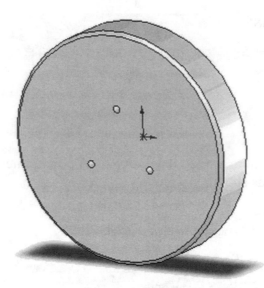

96. Move the cursor to the upper left portion of the screen where the part tree is located. Right click on **Cut-Extrusion1**. A pop up menu will appear. Left click on **Edit Feature** as shown in Figure 91.

Figure 91

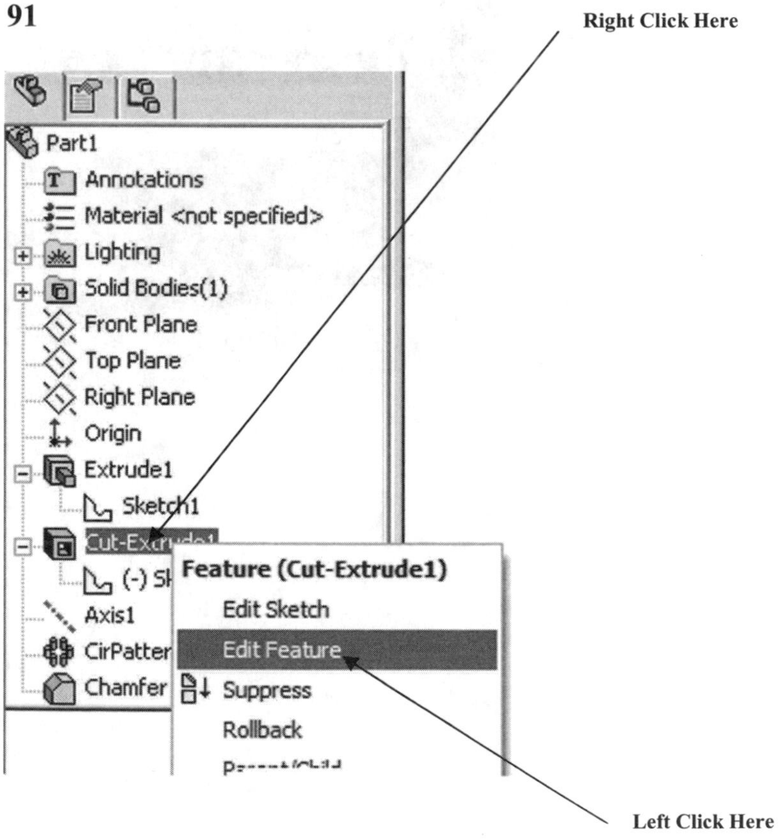

Right Click Here

Left Click Here

97.   Enter **.750** for the extrusion distance as shown in Figure 92.

Figure 92   Enter .750 Here

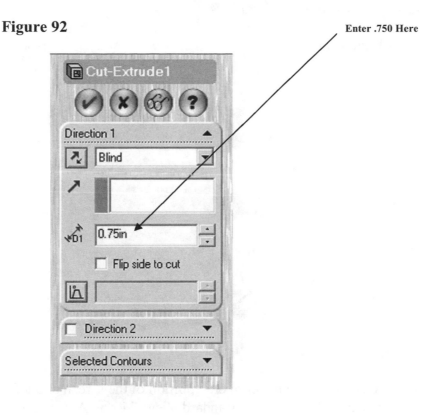

98.   Move the cursor to the upper left portion of the screen and left click on the green checkmark as shown in Figure 93.

Figure 93   Left Click Here

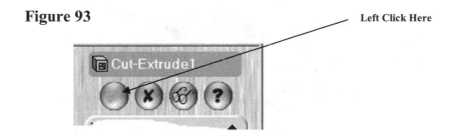

99. SolidWorks will automatically update the part. Notice that the holes are now "thru" holes as shown in Figure 94.

Figure 94

Thru Holes

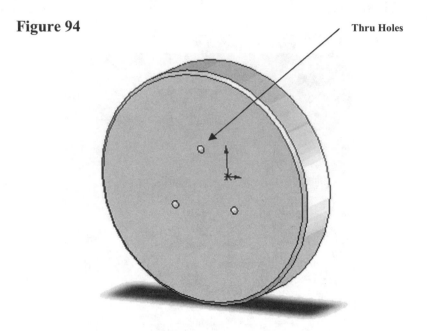

100. Move the cursor to the upper middle portion of the screen and left click on the drop down arrow next to the "Standard Views" icon. A drop down menu will appear. Left click on **Front** as shown in Figure 95.

Figure 95

Left Click Here

Left Click Here

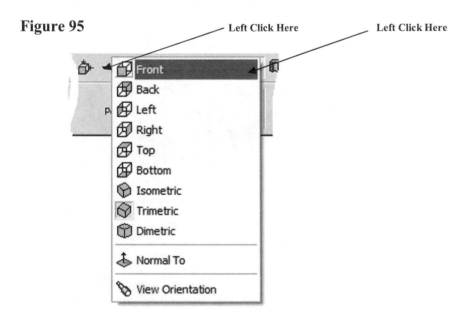

101. A perpendicular view of the part will be displayed. Verify the holes are actually thru holes as shown in Figure 96.

Figure 96

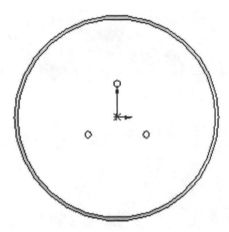

102. Move the cursor to the upper left portion of the screen where the part tree is located. Right click on **CirPattern1**. A pop up menu will appear. Left click on **Edit Feature** as shown in Figure 97.

Figure 97

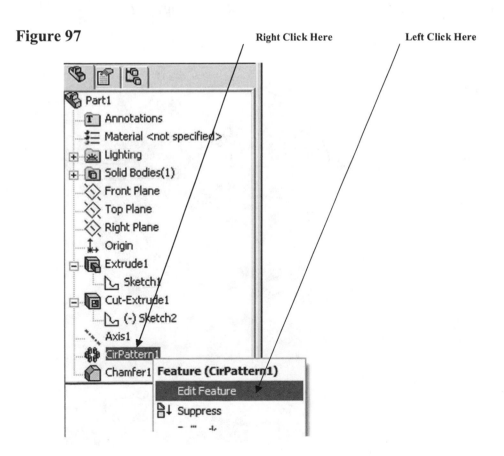

103. Enter **6** for the number of holes and left click next to Equal Spacing as shown in Figure 98.

Figure 98

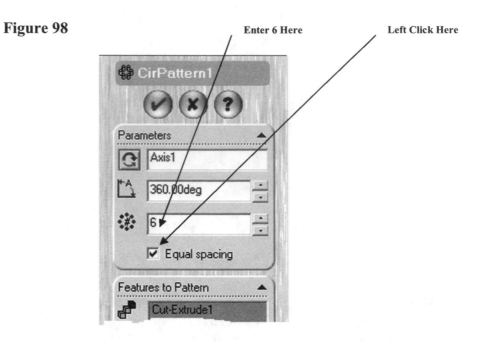

104. A preview will be displayed as shown in Figure 99.

Figure 99

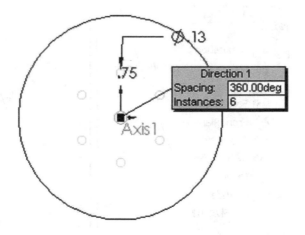

105. Left click on the green checkmark as shown in Figure 100.

Figure 100

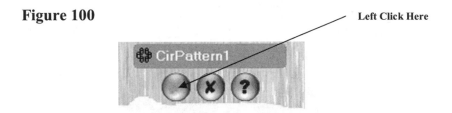

Left Click Here

106. Your screen should look similar to Figure 101.

Figure 101

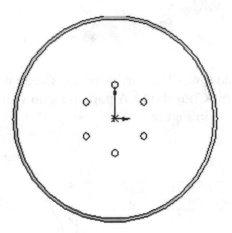

107. Move the cursor to the upper right portion of the screen and left click on the drop down arrow next to the "Standard Views" Icon. A drop down menu will appear. Left click on **Trimetric** as shown in Figure 102.

Figure 102

Left Click Here

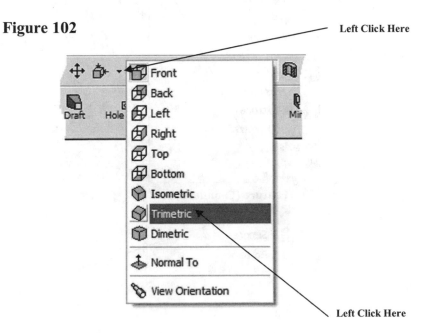

Left Click Here

108. The part will be displayed in trimetric as shown in Figure 103.

Figure 103

109. Move the cursor to the lower left portion of the screen where the part tree is located. Right click on **Chamfer1**. A pop up menu will appear. Left click on **Edit Feature** as shown in Figure 104.

Figure 104

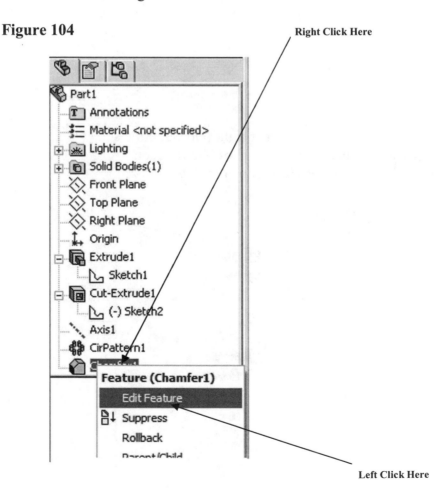

110. Enter **.25** for the Distance as shown in Figure 105.

Figure 105

Enter .25 Here

111. A preview of the chamfer will be displayed as shown in Figure 106.

Figure 106

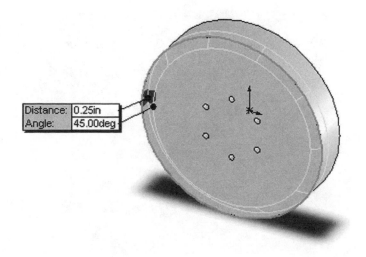

112.   Left click on the green checkmark as shown in Figure 107.

   Figure 107

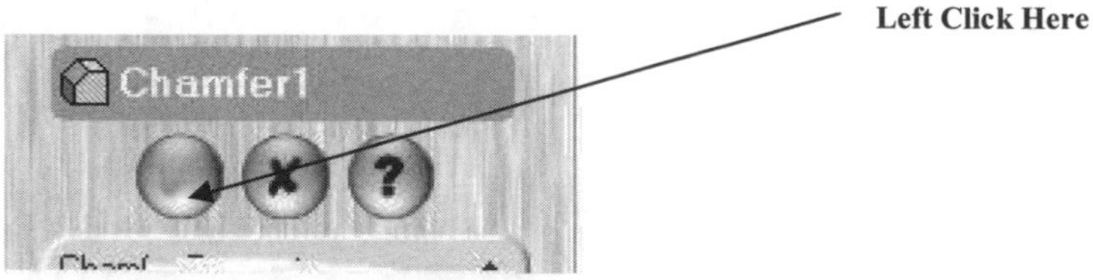

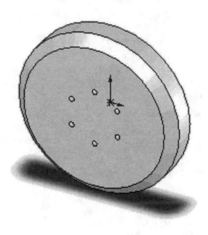

Left Click Here

113.   Your screen should look similar to Figure 108.

   Figure 108

114. Move the cursor to the middle left portion of the screen where the part tree is located. Right click on **CirPattern1.** A pop up menu will appear. Left click on **Suppress** as shown in Figure 109.

Figure 109

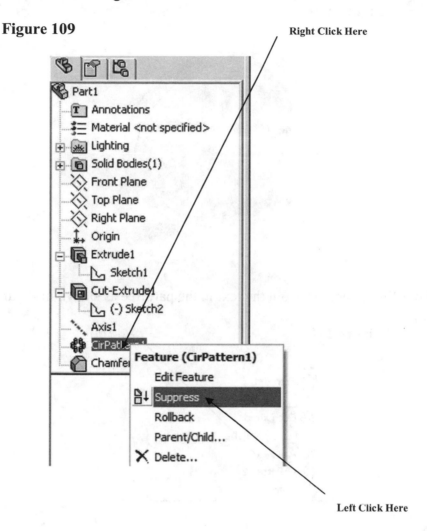

251

115. All holes created using the CirPattern command will be suppressed except for the original hole as shown in Figure 110.

Figure 110

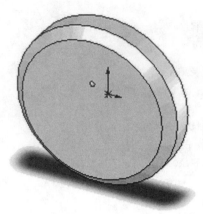

116. SolidWorks will gray out the text in the part tree as shown in Figure 111.

Figure 111
Grayed Out

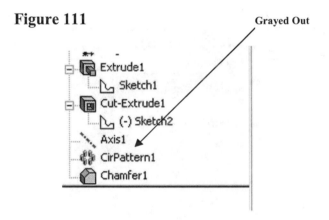

117. The names of all branches in the part tree can also be edited. Left click once on the text **Extrude1**. The text will become highlighted as shown in Figure 112. After the text is highlighted, left click once. The text can now be edited.

Figure 112

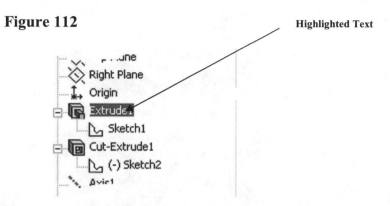

Highlighted Text

118. Enter the text **Base Extrusion** as shown in Figure 113. Press **Enter** on the keyboard. Text for each of the individual operations can be edited if desired.

Figure 113

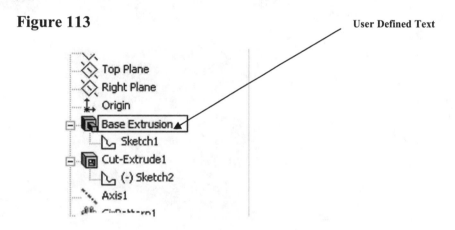

User Defined Text

119. Notice that the final design looks significantly different than the original design. The new part was redesigned by modifying the existing part as shown in Figure 114.

Figure 114

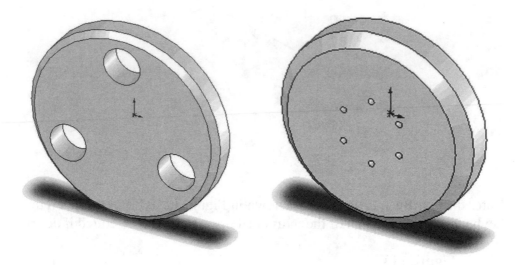

Drawing Activities

Use these problems from Chapters 1 and 2 to create redesigned parts.

Problem 1

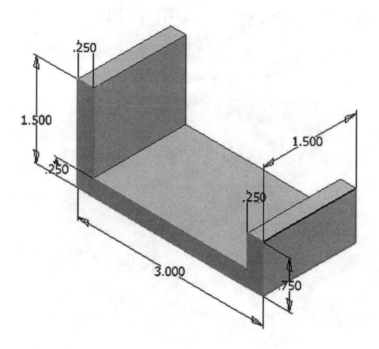

Problem 2

Extrude Center Section .25 Deep

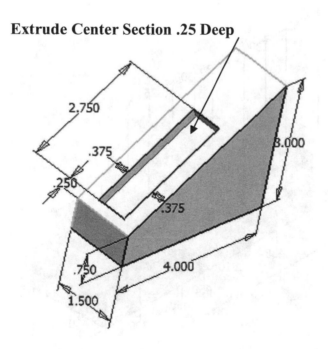

Problem 3

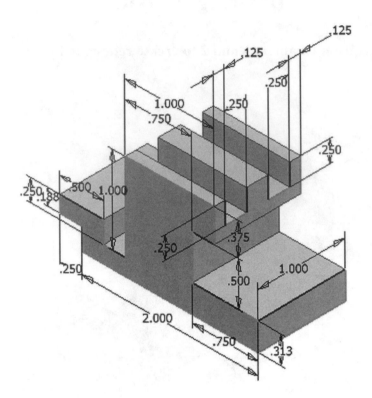

Problem 4

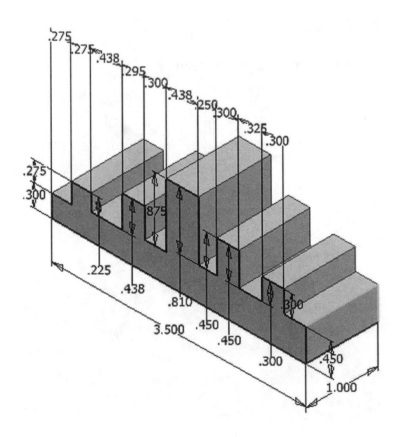

Problem 5

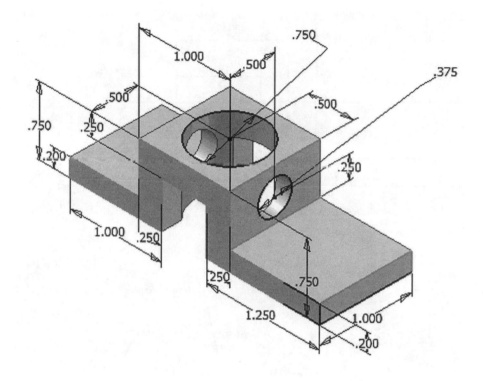

Problem 6

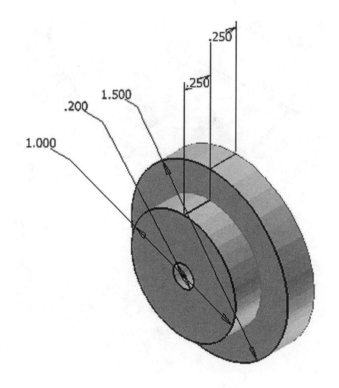

Problem 7

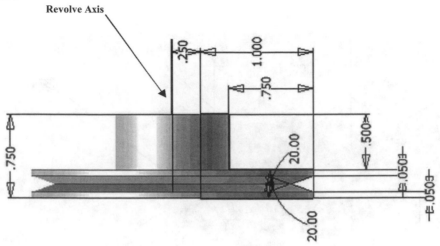

Revolve Axis

Problem 8

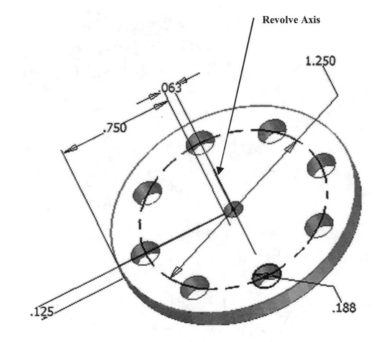

Revolve Axis

Chapter 6 Advanced Design Procedures

Objectives:

- Design multiple sketch parts
- Learn to use the Front, Top, and Right Planes
- Learn to use the Shell command
- Learn to use the Wireframe viewing command

Chapter 6 includes instruction on how to design the parts shown below.

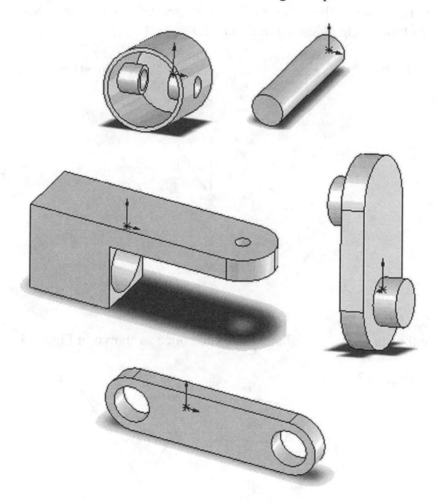

1.   Start SolidWorks by referring to "Chapter 1 Getting Started".

2.   After SolidWorks is running, begin a new sketch.

3.   Move the cursor to the upper middle portion of the screen and left click on **Circle** as shown in Figure 1.

Figure 1   Left Click Here

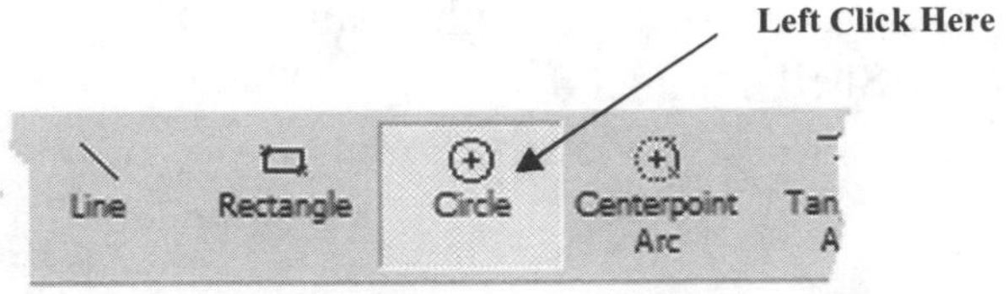

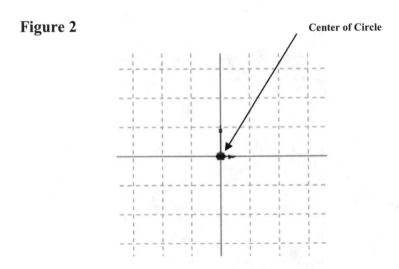

4.   Move the cursor to the center of the screen and left click once. This will be the center of the circle as shown in Figure 2.

Figure 2   Center of Circle

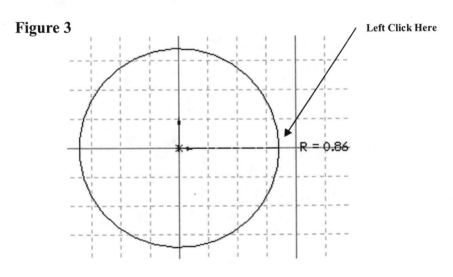

5.   Move the cursor to the right and left click once as shown in Figure 3.

Figure 3   Left Click Here

R = 0.86

6. Right click anywhere around the drawing. A pop up menu will appear. Left click on **Select** as shown in Figure 4.

Figure 4

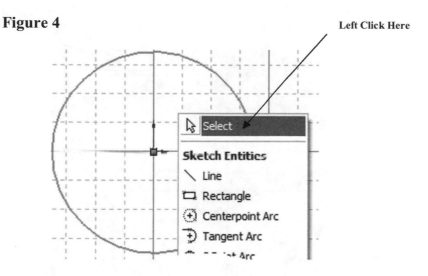

Left Click Here

7. Move the cursor to the upper left portion of the screen and left click on **Smart Dimension** as shown in Figure 5.

Figure 5

Left Click Here

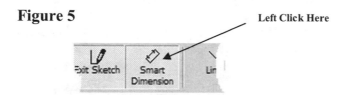

8. After selecting **Smart Dimension** move the cursor over the edge of the circle causing it to turn red as shown in Figure 6. Left click once. The dimension will be attached to the cursor.

Figure 6

Turned Red

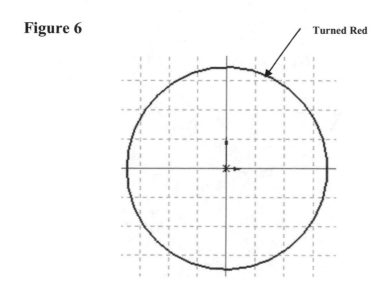

9. Move the cursor to where the dimension will be placed and left click once as shown in Figure 7.

Figure 7

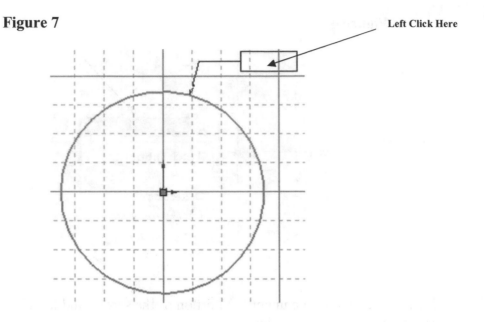

Left Click Here

10. The Modify dialog box will appear as shown in Figure 8.

Figure 8

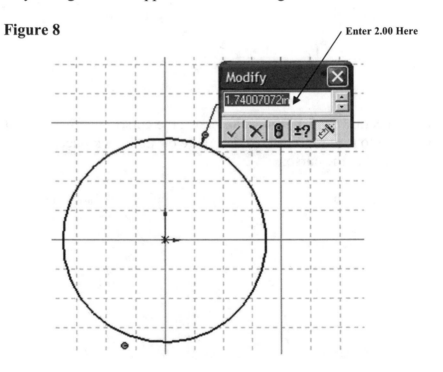

Enter 2.00 Here

11. To edit the dimension, type **2.00** in the Modify dialog box (while the current dimension is highlighted). Left click on the green checkmark as shown in Figure 9.

Figure 9

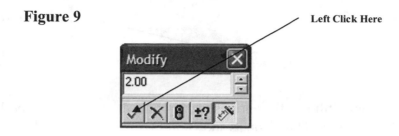

12. The dimension of the circle will become 2.00 inches as shown in Figure 10.

Figure 10

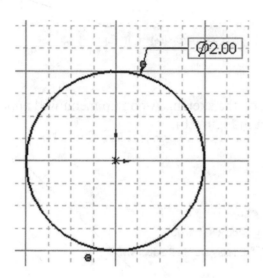

13. Move the cursor to the upper left portion of the screen and left click on the green checkmark as shown in Figure 11.

Figure 11

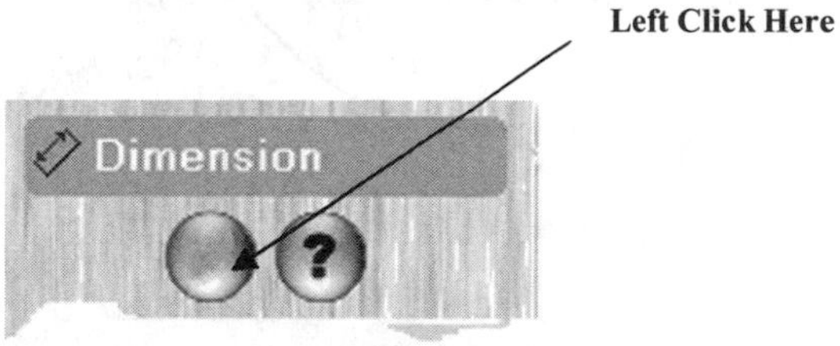

Left Click Here

263

14. In order to view the entire drawing it may be necessary to move the cursor to the middle portion of the screen and left click once on the "Zoom All" icon as shown in Figure 12.

Figure 12

Left Click Here

15. The drawing will "fill up" the entire screen. If the drawing is still too large, left click on the "Zoom In/Out" icon as shown in Figure 13. After selecting the Zoom In/Out icon, hold the left mouse button down and drag the cursor up or down to achieve the desired view of the sketch.

Figure 13

Left Click Here

16. Right click anywhere on the sketch. A pop up menu will appear. Left click on **Exit Sketch** as shown in Figure 14.

Figure 14

Left Click Here

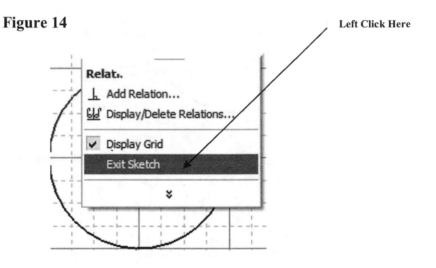

17.   SolidWorks is now out of the Sketch commands and into the Features commands. Notice that the commands at the top of the screen are now different. To work in the Features commands a sketch must be present and have no opens (non-connected lines). If there are any opens in the sketch an error message will appear. Your screen should look similar to Figure 15.

Figure 15

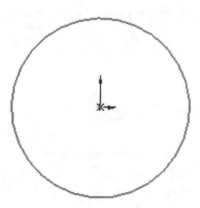

18.   Move the cursor to the upper right portion of the screen and left click on the drop down arrow next to the "Standard Views" icon. A drop down menu will appear. Left click on **Trimetric** as shown in Figure 16.

Figure 16

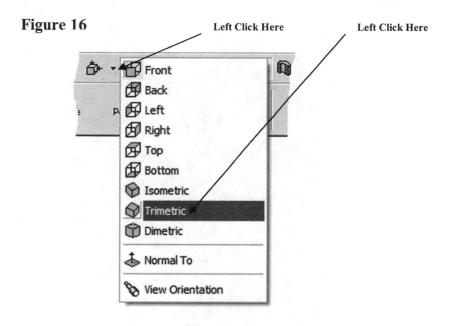

19.   The view will become trimetric as shown in Figure 17.

Figure 17

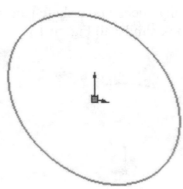

20.   Move the cursor to the upper left portion of the screen and left click on **Extruded Boss/Base**. A preview of the extrusion will be displayed. If SolidWorks gave you an error message, there are opens (non-connected lines) somewhere on the sketch.

21.   Enter **2.00** for D1 and left click on the green checkmark as shown in Figure 18.

Figure 18

Enter 2.00 Here

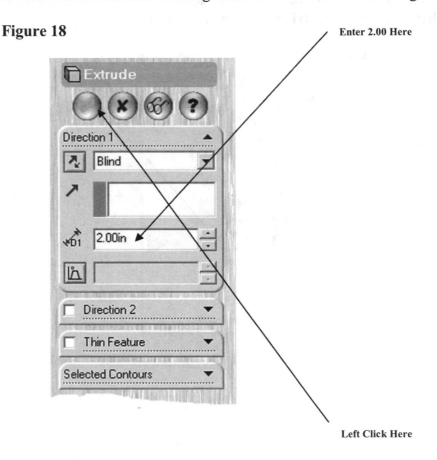

Left Click Here

22.   Your screen should look similar to Figure 19.

Figure 19

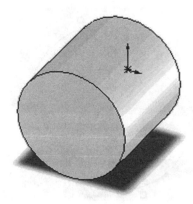

23.   Move the cursor to the upper left portion of the screen causing the text "Right Plane" to become highlighted as shown in Figure 20.

Figure 20   Text Highlighted

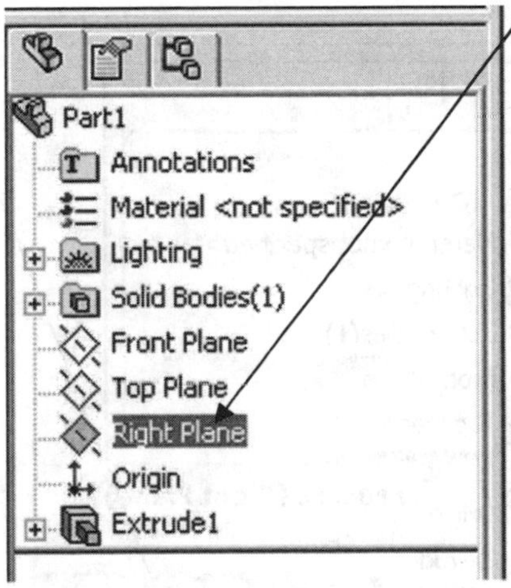

24. The Right Plane will become visible as shown in Figure 21.

Figure 21

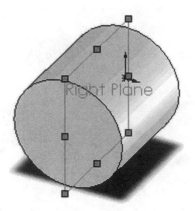

25. Right click on the **Right Plane** text. A pop up menu will appear. Left click on **Insert Sketch** as shown in Figure 22.

Figure 22

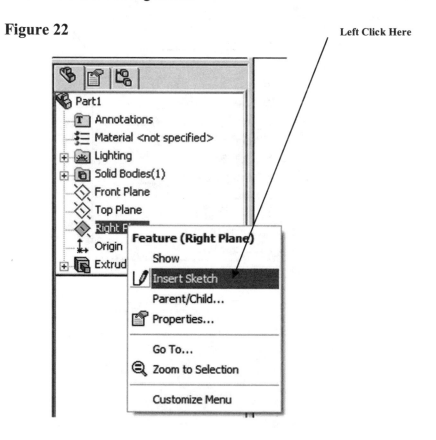

26.   Your screen should look similar to Figure 23.

Figure 23

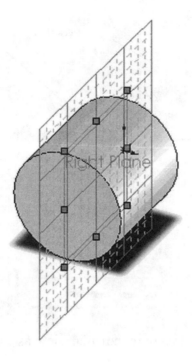

27.   Move the cursor to the upper right portion of the screen and left click on the "Wire Frame Display" icon as shown in Figure 24.

Figure 24   **Left Click Here**

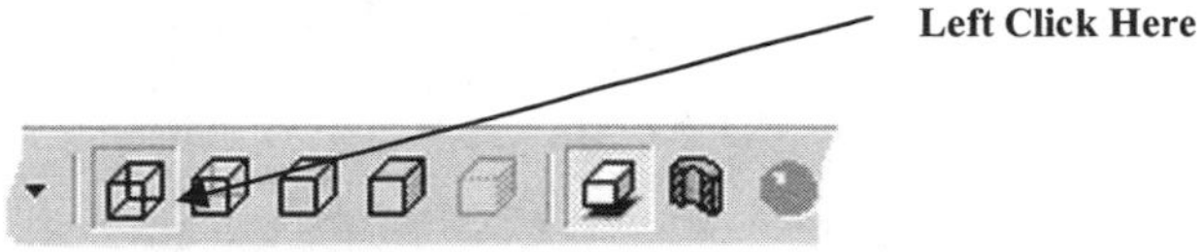

28.   Your screen should look similar to Figure 25.

Figure 25

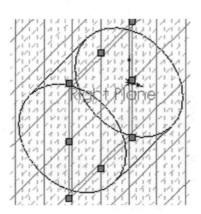

29. Move the cursor to the upper right portion of the screen and left click on the drop down arrow next to the "Standard Views" icon. A drop down menu will appear. Left click on **Right** as shown in Figure 26.

Figure 26

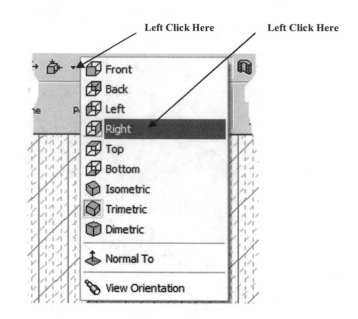

30. The Right Plane will rotate to a perpendicular view as shown in Figure 27.

Figure 27

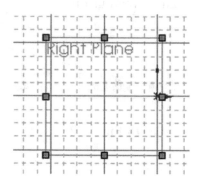

31. The lines that make up the part may not be visible. Move the cursor around where they were located. The lines will appear in red. Move the cursor over the top line causing it to turn red and left click once as shown in Figure 28.

Figure 28

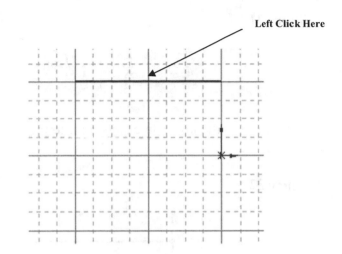

32. While the line is still red move the cursor to the upper right portion of the screen and left click on **Convert Entities** as shown in Figure 29. The line will become visible.

Figure 29

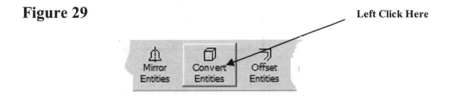

33. Move the cursor over the left vertical line causing it to turn red. Left click once as shown in Figure 30.

Figure 30

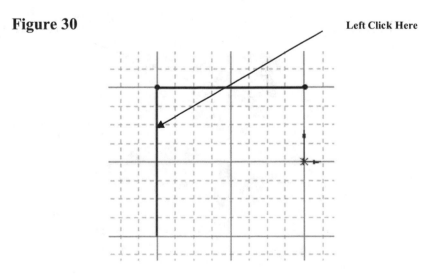

34. While the line is still red move the cursor to the upper right portion of the screen and left click on **Convert Entities** as shown in Figure 31. The line will become visible.

Figure 31

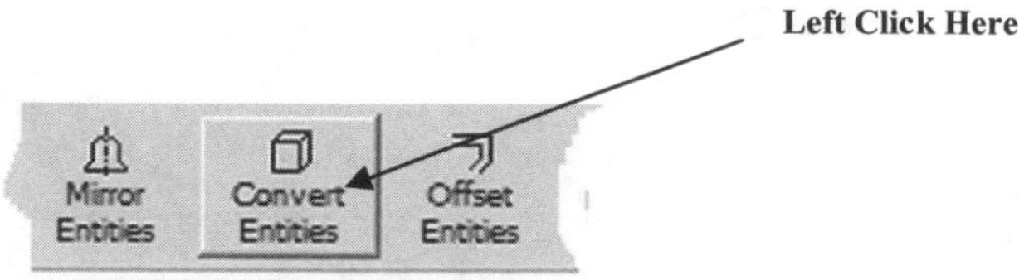

Left Click Here

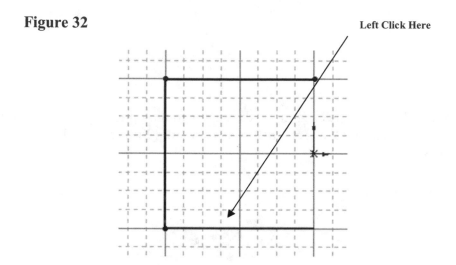

35. Move the cursor over the lower line causing it to turn red. Left click once as shown in Figure 32.

Figure 32

Left Click Here

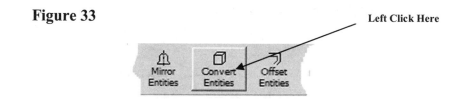

36. While the line is still red move the cursor to the upper right portion of the screen and left click on **Convert Entities** as shown in Figure 33. The line will become visible.

Figure 33

Left Click Here

37.   Move the cursor over the right vertical line causing it to turn red. Left click once as shown in Figure 34.

Figure 34

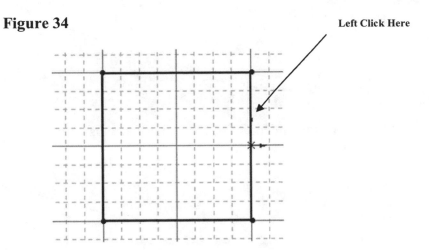

38.   While the line is still red move the cursor to the upper right portion of the screen and left click on **Convert Entities** as shown in Figure 35. The line will become visible.

Figure 35

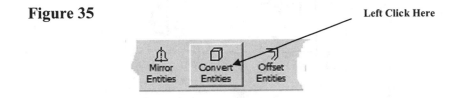

39.   Your screen should look similar to Figure 36.

Figure 36

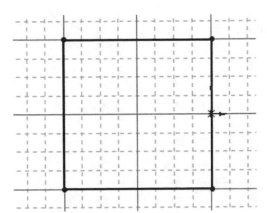

40.   Move the cursor to the upper left portion of the screen and left click on **Line** as shown in Figure 37.

Figure 37

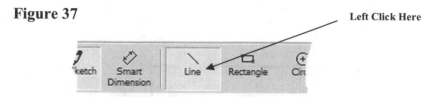

Left Click Here

41.   Move the cursor to the midpoint of the upper line causing a red dot to appear. Left click as shown in Figure 38.

Figure 38   Left Click Here

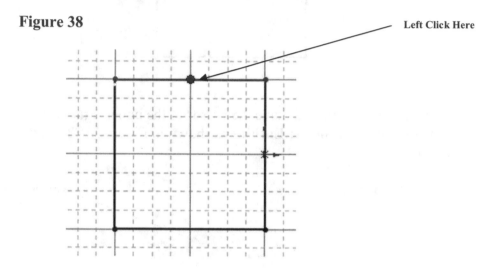

42.   Move the cursor down to the lower line causing a red dot to appear. Left click as shown in Figure 39.

Figure 39   Left Click Here

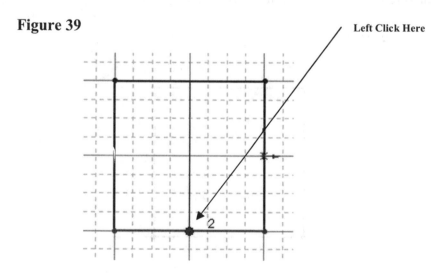

43. Right click anywhere around the drawing. A pop up menu will appear. Left click on **Select** as shown in Figure 40.

Figure 40

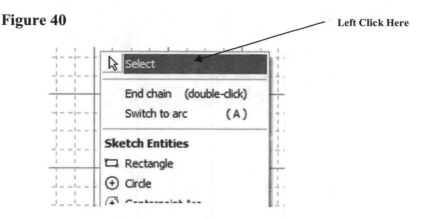

Left Click Here

44. Your screen should look similar to Figure 41.

Figure 41

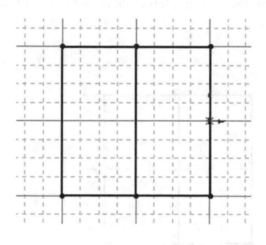

45. Move the cursor to the upper middle portion of the screen and left click on **Circle** as shown in Figure 42.

Figure 42

Left Click Here

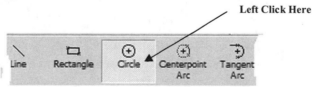

46.   Move the cursor to the midpoint of the center line. A yellow dot will appear. Left click as shown in Figure 43.

Figure 43

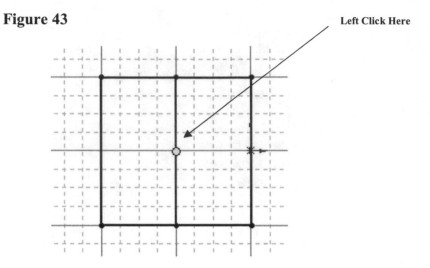

Left Click Here

47.   Move the cursor out to the side and left click as shown in Figure 44.

Figure 44

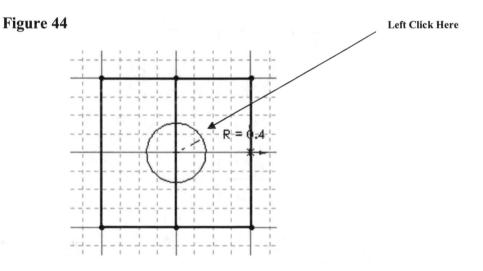

Left Click Here

48. Right click anywhere around the drawing. A pop up menu will appear. Left click on **Select** as shown in Figure 45.

Figure 45

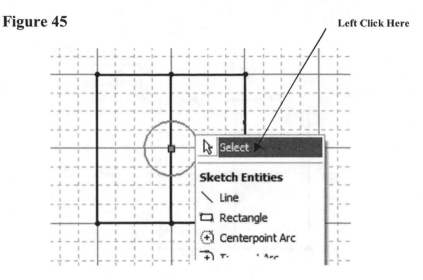

49. Move the cursor to the upper left portion of the screen and left click on **Smart Dimension** as shown in Figure 46.

Figure 46

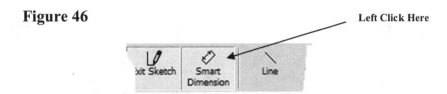

50. Move the cursor over the edge (not center) of the circle until it turns red. Left click once as shown in Figure 47.

Figure 47

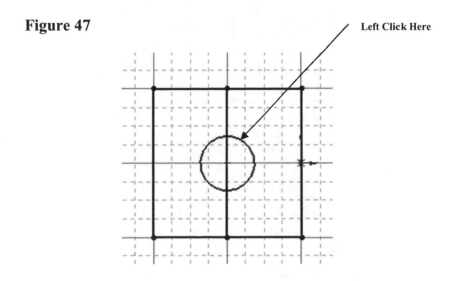

51. Move the cursor around. The dimension of the line will appear as shown in Figure 48. The dimension is attached to the cursor.

Figure 48

Left Click Here

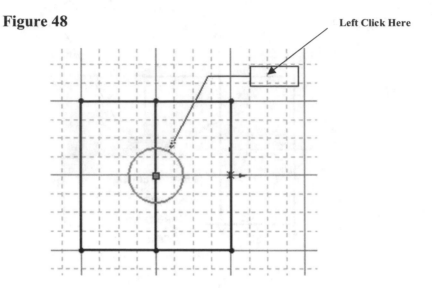

52. Move the cursor to where the dimension will be placed and left click once. The Modify dialog box will appear as shown in Figure 49.

Figure 49

Enter .500 Here

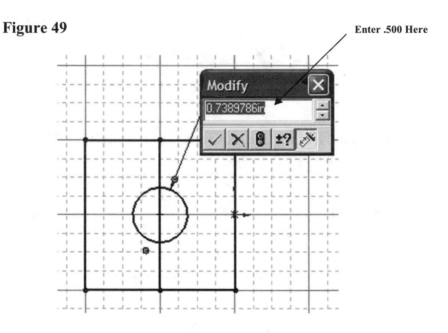

53. To edit the dimension, type **.500** in the Modify dialog box (while the current dimension is highlighted) and left click on the green checkmark as shown in Figure 50.

Figure 50

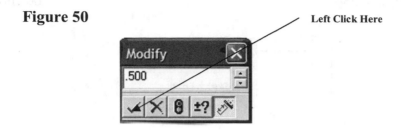

Left Click Here

54. Your screen should look similar to Figure 51.

Figure 51

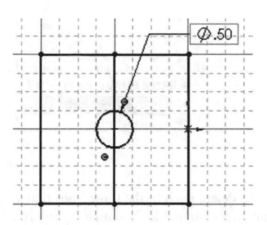

55. Right click anywhere around the drawing. A pop up menu will appear. Left click on **Select** as shown in Figure 52.

Figure 52

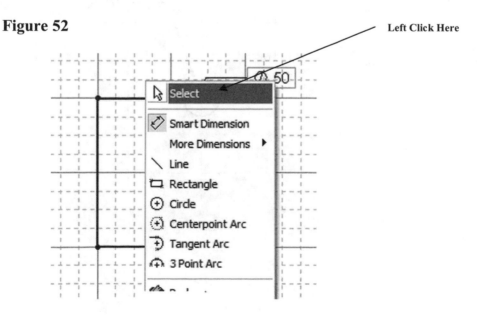

Left Click Here

279

56. Move the cursor over the center line causing it to turn red. Right click on the line. A pop up menu will appear. Left click on **Delete** as shown in Figure 53.

Figure 53

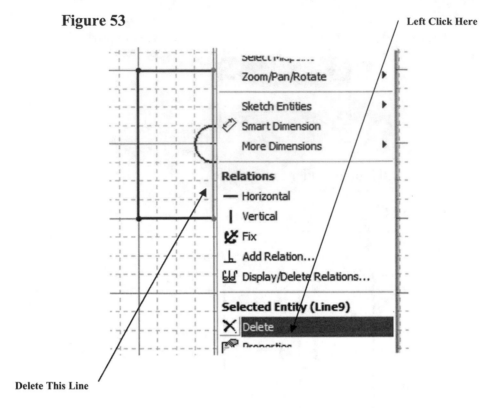

57. Repeat the same steps to delete each of the three lines that were projected on to the Right Plane. Your screen should look similar to Figure 54.

Figure 54

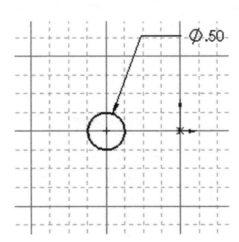

58. Move the cursor to the upper right portion of the screen and left click on the "Shaded with Edges" icon as shown in Figure 55.

Figure 55

Left Click Here

59. Your screen should look similar to Figure 56.

Figure 56

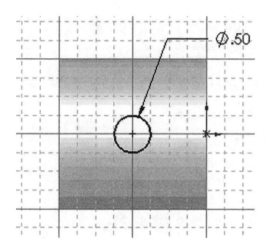

Ø.50

60. After you have verified that no commands are active, right click anywhere on the sketch. A pop up menu will appear. Left click on **Exit Sketch** as shown in Figure 57.

Figure 57

Left Click Here

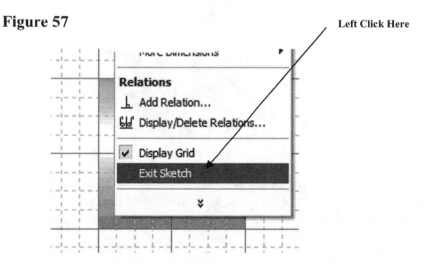

61. Move the cursor to the upper right portion of the screen and left click on the drop down arrow next to the "Standard Views" icon. A drop down menu will appear. Left click on **Trimetric** as shown in Figure 58.

Figure 58

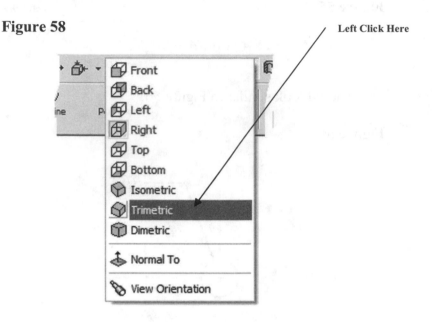

Left Click Here

62. Your screen should look similar to Figure 59.

Figure 59

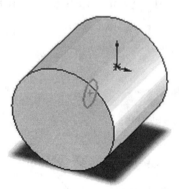

63. Move the cursor to the upper left portion of the screen and left click on **Extruded Cut** as shown in Figure 60. The Extrude dialog box will appear. Move the cursor inside the small circle causing it to turn red. Left click once.

Figure 60

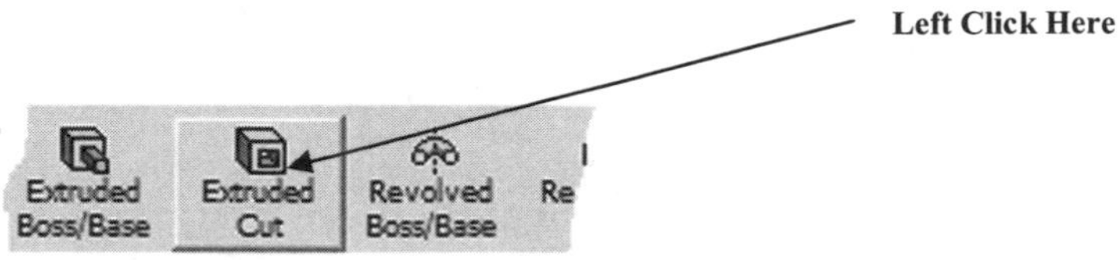

Left Click Here

64. Move the cursor to the middle left portion of the screen. Enter **1.00** for D1. Left click in the box next to the text "Direction 2". Enter **1.00** for D2 as shown in Figure 61.

Figure 61

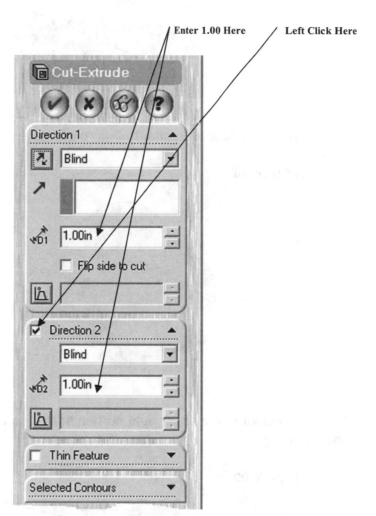

Enter 1.00 Here   Left Click Here

65. A preview will be displayed as shown in Figure 62.

Figure 62

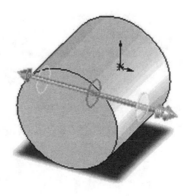

66. Move the cursor to the upper left portion of the screen and left click on the green checkmark as shown in Figure 63.

Figure 63

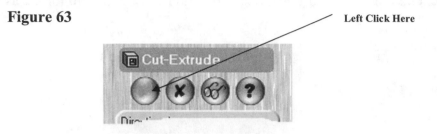

Left Click Here

67. Your screen should look similar to Figure 64.

Figure 64

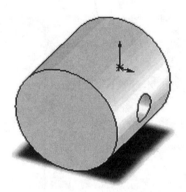

68. Move the cursor to the upper right portion of the screen and left click on **Shell** as shown in Figure 65.

Figure 65

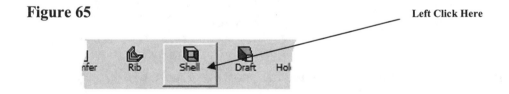

Left Click Here

69. Left click on the lower surface as shown in Figure 66.

Figure 66

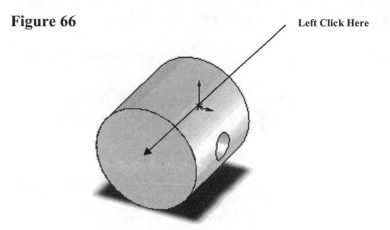

Left Click Here

70. Enter **.10** next to D1. Left click on the green checkmark as shown in Figure 67.

Figure 67

Enter .10 Here   Left Click Here

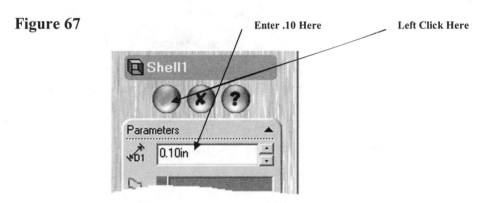

71. Your screen should look similar to Figure 68.

Figure 68

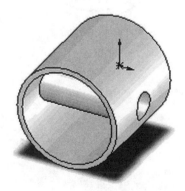

72. Move the cursor to the upper right portion of the screen and left click on the drop down arrow next to the "Standard Views" icon. A drop down menu will appear. Left click on **Front** as shown in Figure 69.

Figure 69

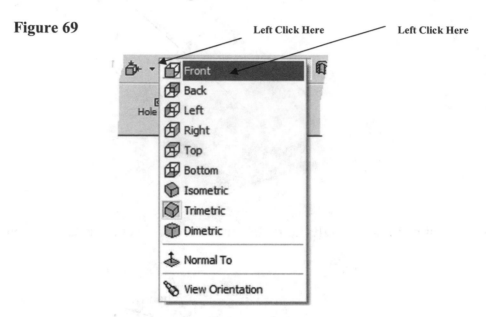

73. The part will rotate to provide a perpendicular view of the inside as shown in Figure 70.

Figure 70

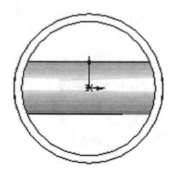

74. Move the cursor over the same surface causing both the inside and outside lines to turn red. You may have to zoom in for SolidWorks to find both lines at the same time. Both lines must be red at the same time. After both lines are red at the same time, right click on the surface as shown in Figure 71. The surface will turn green.

Figure 71

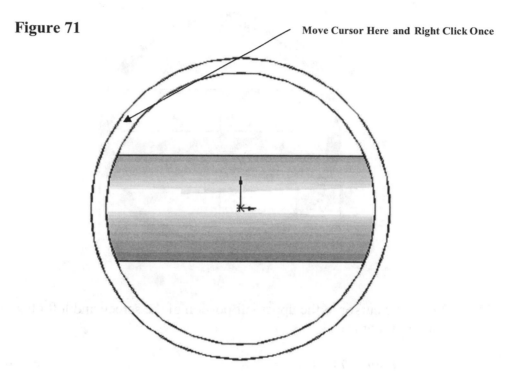

Move Cursor Here and Right Click Once

75. A pop up menu will appear. Left click on **Insert Sketch** as shown in Figure 72.

Figure 72

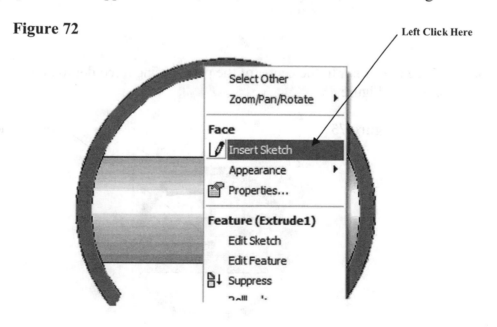

Left Click Here

76. A new sketch will appear on the selected surface. Your screen should look similar to Figure 73.

Figure 73

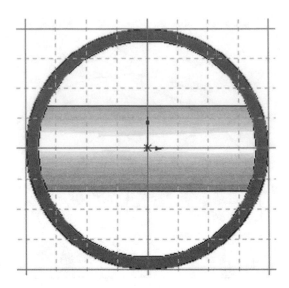

77. Move the cursor to the upper left portion of the screen and left click on **Line** as shown in Figure 74.

Figure 74

Left Click Here

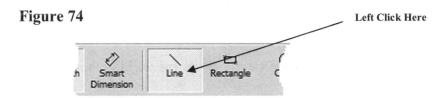

78. Move the cursor to the center of the part causing a red dot to appear. Left click as shown in Figure 75.

Figure 75

Left Click Here

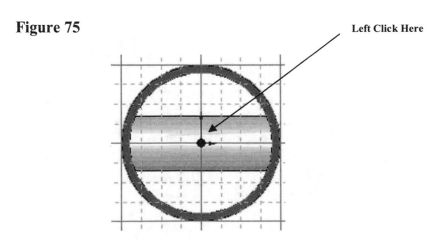

79. Move the cursor upward and left click as shown in Figure 76.

Figure 76

Left Click Here

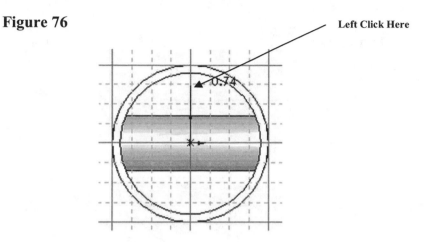

80. Right click anywhere around the drawing. A pop up menu will appear. Left click on **Select** as shown in Figure 77.

Figure 77

Left Click Here

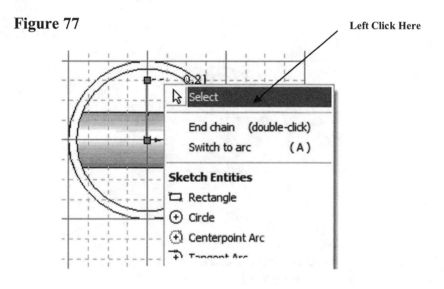

81. Move the cursor to the upper left portion of the screen and left click on **Line** as shown in Figure 78.

Figure 78

Left Click Here

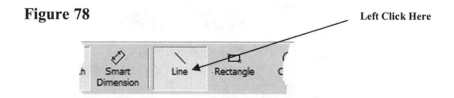

82. Move the cursor to the center of the part causing a red dot to appear. Left click as shown in Figure 79.

Figure 79

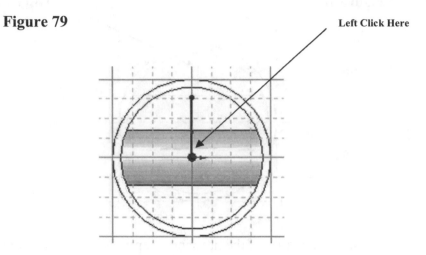

83. Move the cursor to the left and left click as shown in Figure 80.

Figure 80

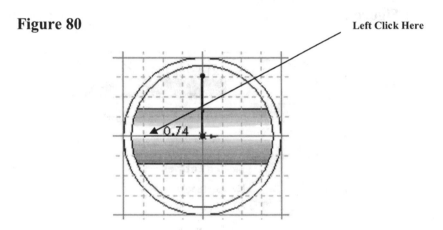

84. Right click anywhere around the drawing. A pop up menu will appear. Left click on **Select** as shown in Figure 81.

Figure 81

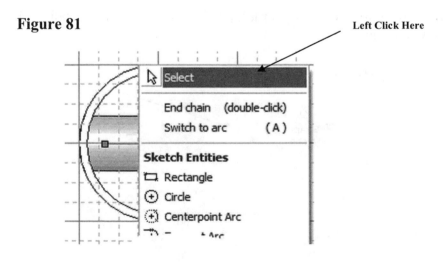

85. Your screen should look similar to Figure 82.

Figure 82

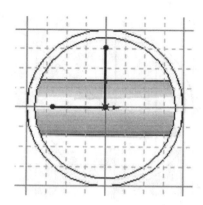

86. Move the cursor to the upper left portion of the screen and left click on **Line** as shown in Figure 83.

Figure 83　　　　　　　　　　　　　　　　　Left Click Here

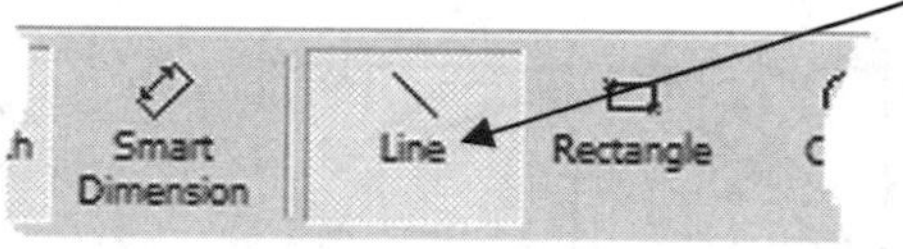

87. Move the cursor to the position shown in Figure 84 and left click once.

Figure 84　　　　　　　　　　　　　　　　Left Click Here

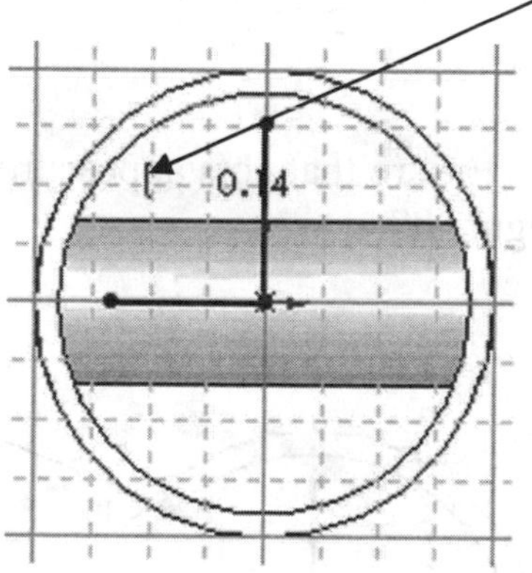

0.14

88. Move the cursor downward and left click as shown in Figure 85.

Figure 85

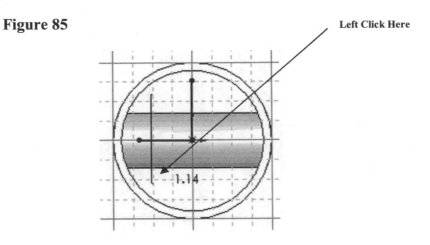

89. Move the cursor to the right and left click as shown in Figure 86.

Figure 86

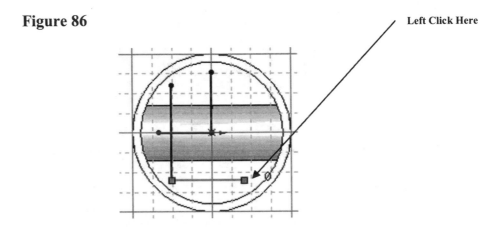

90. Move the cursor upward. Ensure that dots appear from the original starting point. Left click as shown in Figure 87.

Figure 87

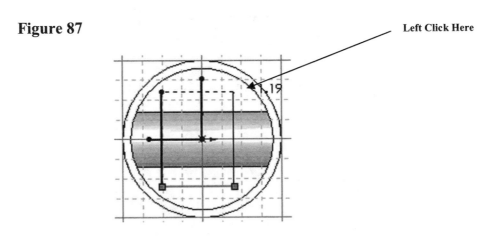

91. Move the cursor to the left and left click as shown in Figure 88.

Figure 88

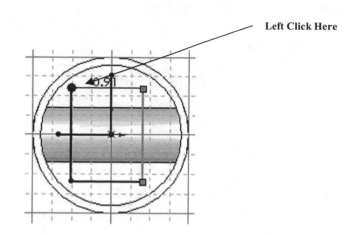

Left Click Here

92. Right click anywhere around the drawing. A pop up menu will appear. Left click on **Select** as shown in Figure 89.

Figure 89

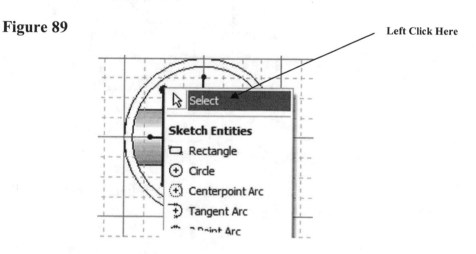

Left Click Here

93. Your screen should look similar to Figure 90.

Figure 90

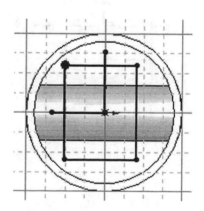

94. Move the cursor to the middle left portion of the screen and left click on **Smart Dimension** as shown in Figure 91.

Figure 91

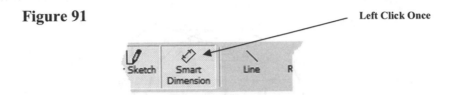

Left Click Once

95. After selecting **Smart Dimension** move the cursor over the lower endpoint of the vertical line coming out of the center of the part. A red dot will appear. Left click on the endpoint as shown in Figure 92.

Figure 92

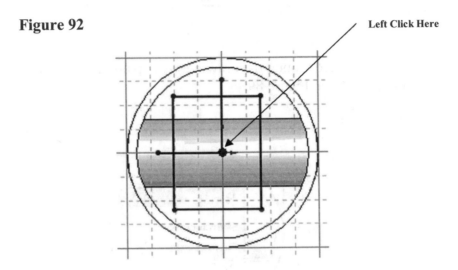

Left Click Here

96. Move the cursor to the endpoint of the far left line. A red dot will appear. Left click once as shown in Figure 93.

Figure 93

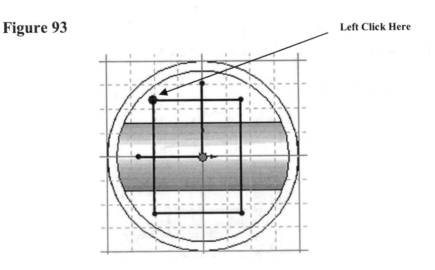

Left Click Here

97. Move the cursor upward. The dimension of the line will appear as shown in Figure 94. The dimension is attached to the cursor.

Figure 94

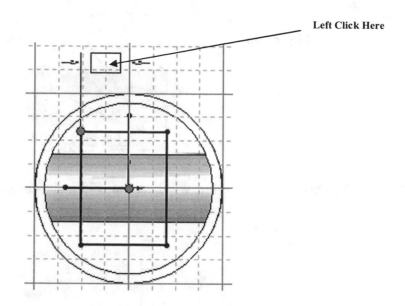

Left Click Here

98. Move the cursor to where the dimension will be placed and left click once. The Modify dialog box will appear as shown in Figure 95.

Figure 95

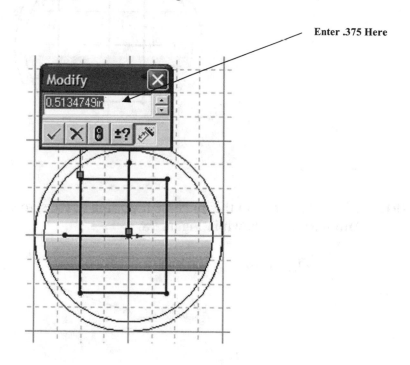

Enter .375 Here

99. To edit the dimension, type **.375** in the Modify dialog box (while the current dimension is highlighted). Left click on the green checkmark as shown in Figure 96.

Figure 96

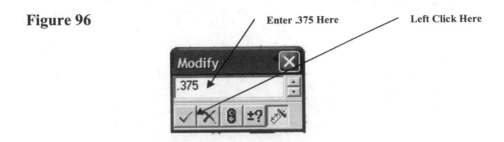

100. The dimension of the line will become .375 inches as shown in Figure 97.

Figure 97

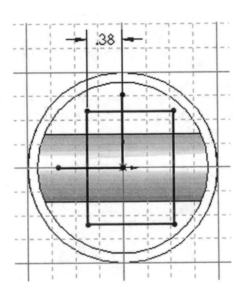

101. Move the cursor to the middle left portion of the screen and left click on **Smart Dimension** as shown in Figure 98.

Figure 98

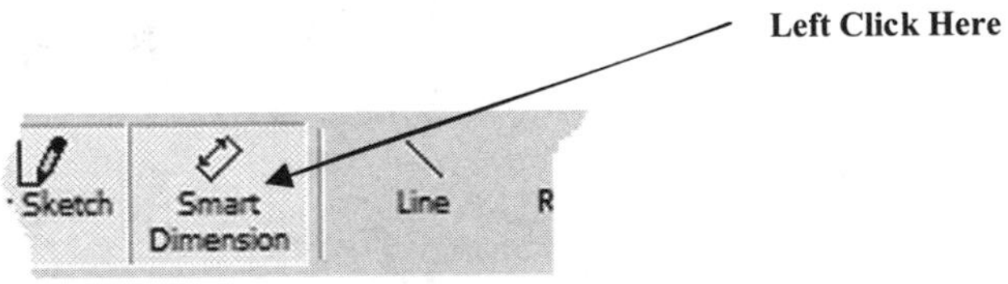

296

102. Move the cursor over the end point of the horizontal line coming out of the center. Left click on the endpoint as shown in Figure 99.

Figure 99

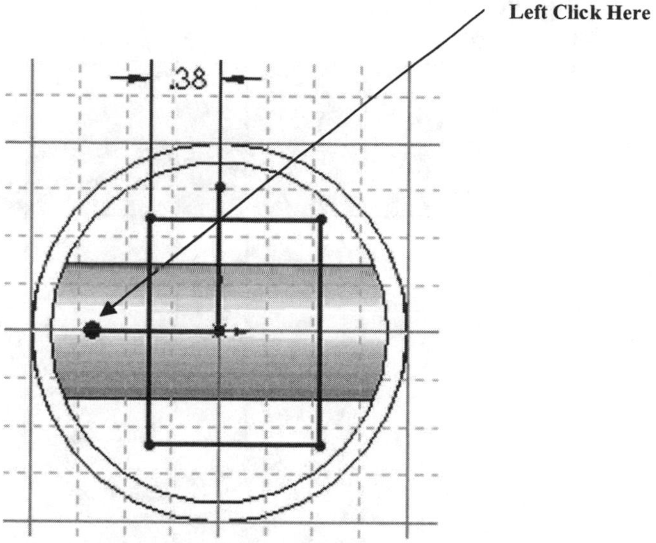

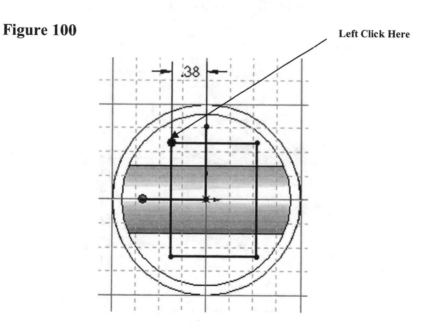

103. Move the cursor to the endpoint of the upper line and left click once as shown in Figure 100.

Figure 100

104. Move the cursor out to the side. The dimension of the line will appear as shown in Figure 101. The dimension is attached to the cursor.

Figure 101

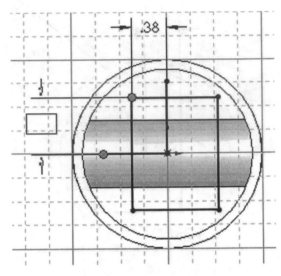

105. Move the cursor to where the dimension will be placed and left click once. The Modify dialog box will appear as shown in Figure 102.

Figure 102

Enter .375 Here

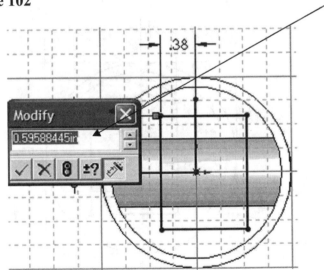

106. To edit the dimension, type **.500** in the Modify dialog box (while the current dimension is highlighted). Left click on the green checkmark as shown in Figure 103.

Figure 103

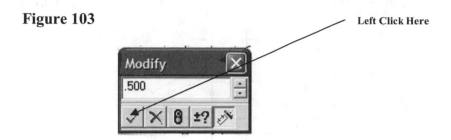

Left Click Here

107. The dimension of the line will become .500 inches as shown in Figure 104.

Figure 104

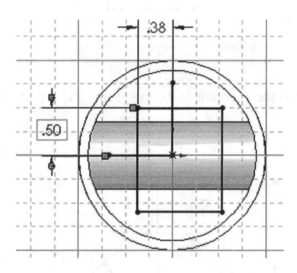

108. Move the cursor to the middle left portion of the screen and left click on **Smart Dimension** as shown in Figure 105.

Figure 105

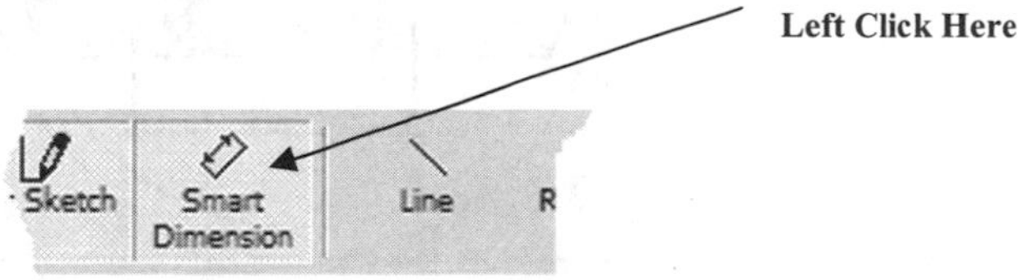

Left Click Here

299

109. Move the cursor over the endpoint of the horizontal line coming out of the center of the part. Left click on the endpoint as shown in Figure 106.

Figure 106

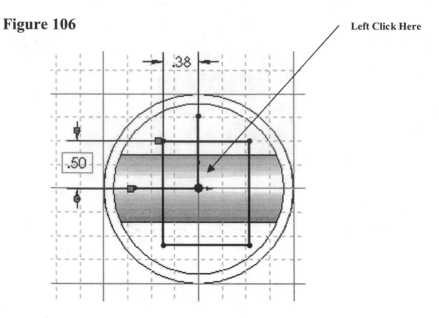

110. Move the cursor to the endpoint of the lower line and left click once as shown in Figure 107.

Figure 107

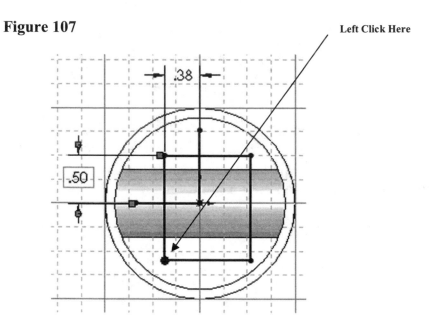

111. Move the cursor out to the side. The dimension of the line will appear as shown in Figure 108. The dimension is attached to the cursor.

Figure 108

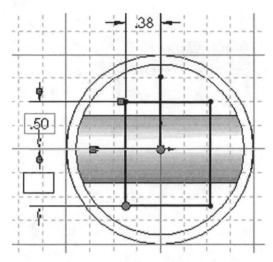

112. Move the cursor to where the dimension will be placed and left click once. The Modify dialog box will appear as shown in Figure 109.

Figure 109

Enter .375 Here

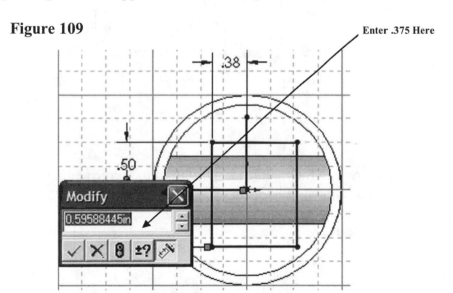

113. To edit the dimension, type **.500** in the Modify dialog box (while the current dimension is highlighted). Left click on the green checkmark as shown in Figure 110.

Figure 110

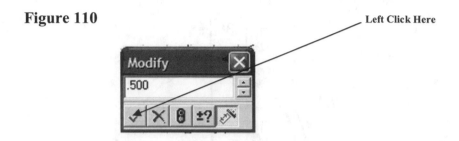

114. The dimension of the line will become .500 inches as shown in Figure 111.

Figure 111

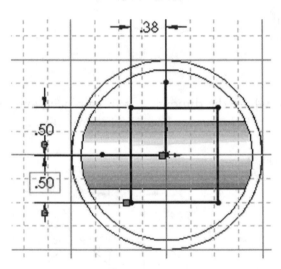

115. Move the cursor to the middle left portion of the screen and left click on **Smart Dimension** as shown in Figure 112.

Figure 112

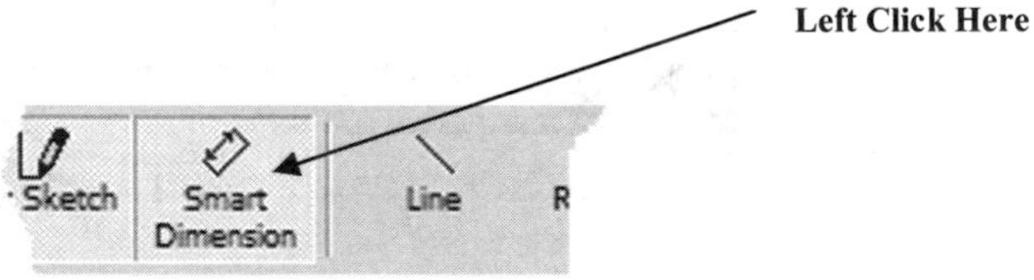

116. After selecting **Smart Dimension** move the cursor to the upper endpoint of the vertical line coming out of the center of the part. Left click as shown in Figure 113.

Figure 113

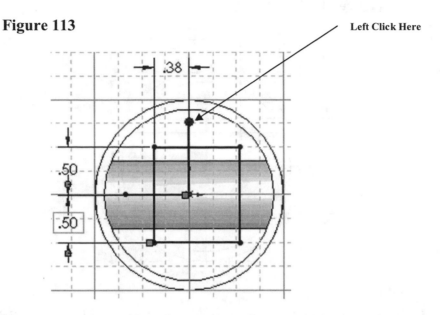

117. Move the cursor to the endpoint of the right vertical line and left click once as shown in Figure 114.

Figure 114

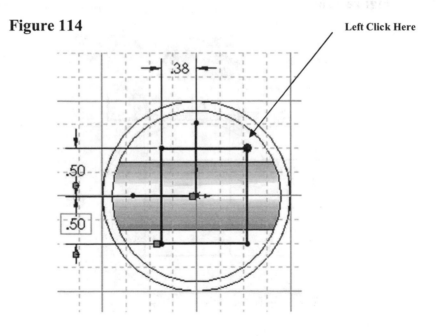

118. Move the cursor out to the side. The dimension of the line will appear as shown in Figure 115. The dimension is attached to the cursor.

Figure 115

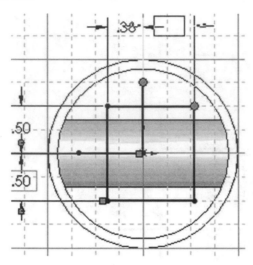

119. Move the cursor to where the dimension will be placed and left click once. The Modify dialog box will appear as shown in Figure 116.

Figure 116

Enter .375 Here

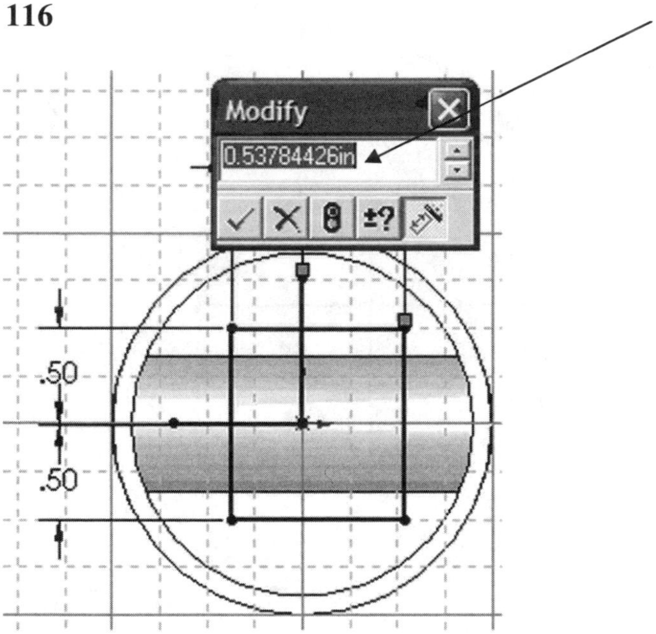

120. To edit the dimension, type **.375** in the Modify dialog box (while the current dimension is highlighted) and left click on the green checkmark as shown in Figure 117.

Figure 117

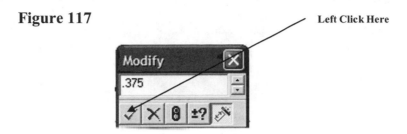

Left Click Here

121. The dimension of the line will become .375 inches as shown in Figure 118.

Figure 118

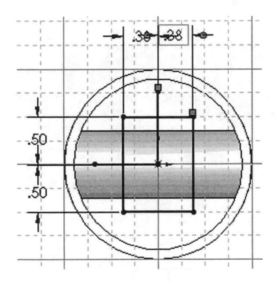

122. Right click anywhere around the drawing. A pop up menu will appear. Left click on **Select** as shown in Figure 119.

Figure 119

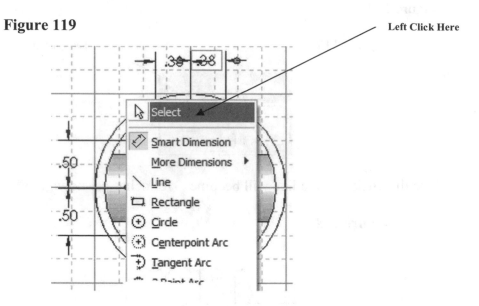

Left Click Here

123. Move the cursor over the vertical line coming out of the center of the part causing it to turn red as shown in Figure 120.

Figure 120

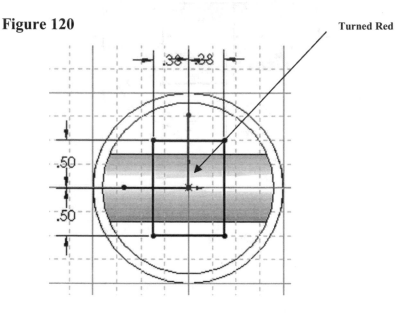

Turned Red

124. After the line has turned red, right click once. A pop up menu will appear. Left click on **Delete** as shown in Figure 121.

Figure 121

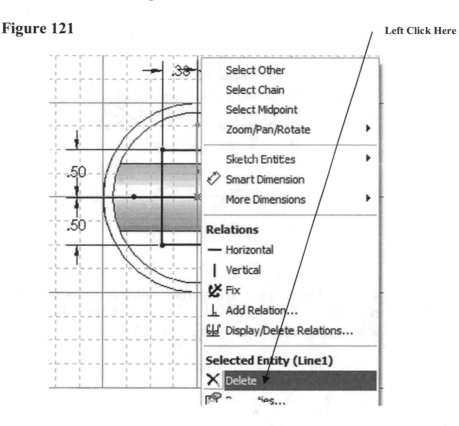

125. Use the same steps to delete the horizontal line coming out of the center of the part as shown in Figure 122. If the dimensions disappear do not be concerned.

Figure 122

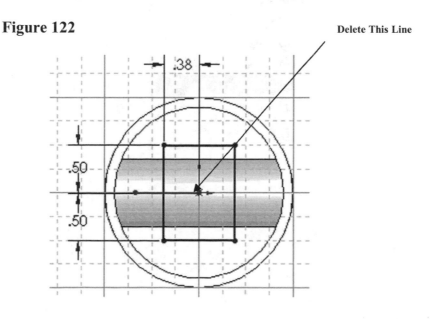

126. Right click anywhere around the drawing. A pop up menu will appear. Left click on **Exit Sketch** as shown in Figure 123.

Figure 123

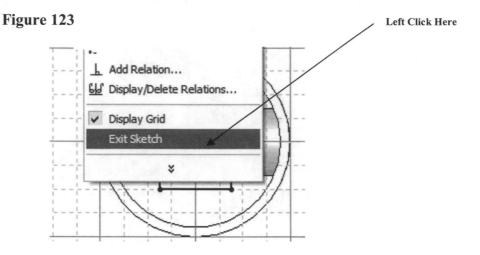

Left Click Here

127. Move the cursor to the upper right portion of the screen and left click on the drop down arrow next to the "Standard Views" icon. A drop down menu will appear. Left click on **Trimetric** as shown in Figure 124.

Figure 124

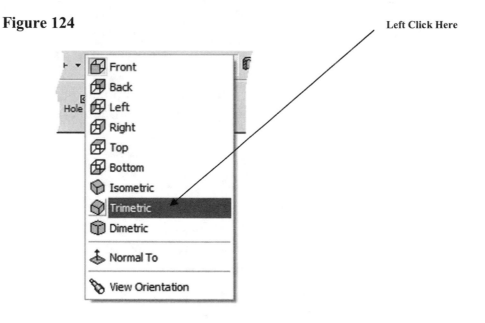

Left Click Here

128. Your screen should look similar to Figure 125.

Figure 125

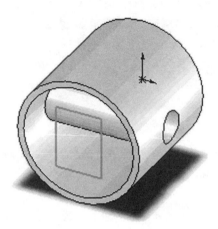

129. Move the cursor to the upper left portion of the screen and left click on **Extruded Cut** as shown in Figure 126.

Figure 126   Left Click Here

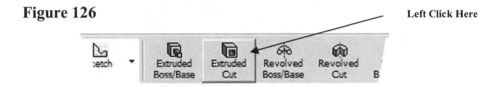

130. Enter **1.875** for D1 as shown in Figure 127.

Figure 127   Enter 1.875 Here

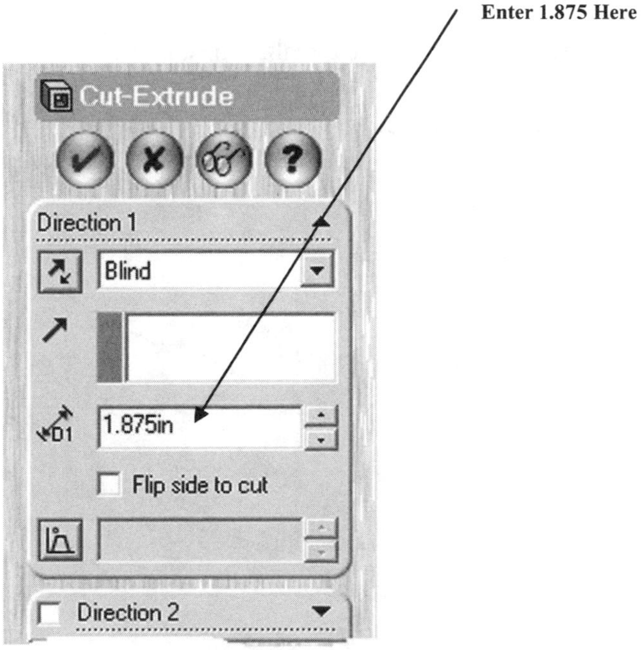

131. Left click on the green checkmark as shown in Figure 128.

Figure 128

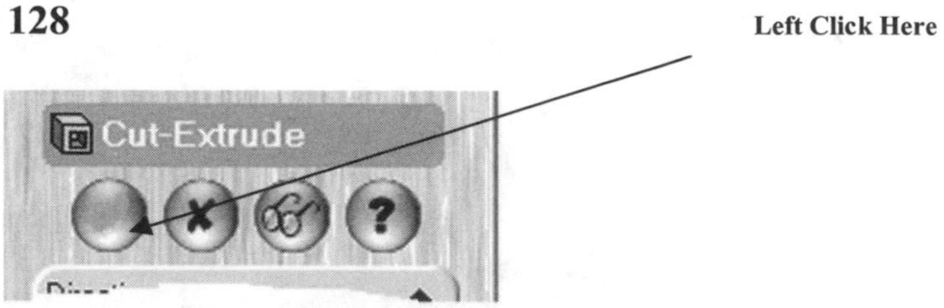

Left Click Here

132. Your screen should look similar to Figure 129. Use the Rotate command to roll the part around to view the inside.

Figure 129

133. Save the part as Piston1.SLDPRT where it can be easily retired later.

134. Begin a new drawing as described in Chapter 1.

135. Draw a circle in the center of the grid as shown in Figure 130.

Figure 130

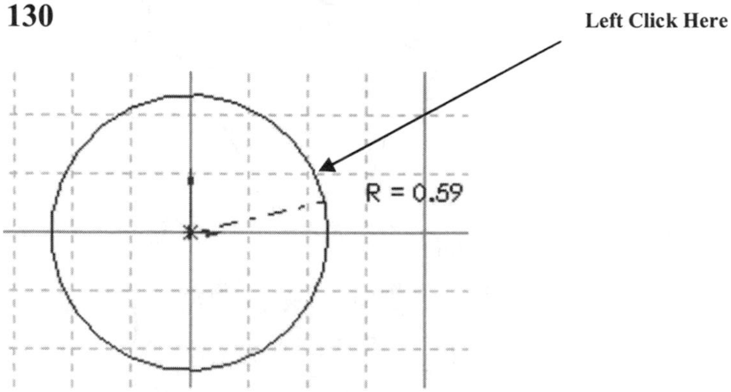

Left Click Here

R = 0.59

136. Use the **Smart Dimension** command to dimension the circle to **.500** inches as shown in Figure 131.

Figure 131

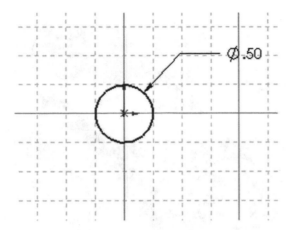

137. Exit the Sketch commands and Extrude the circle to a length of **1.875** inches as shown in Figure 132.

Figure 132

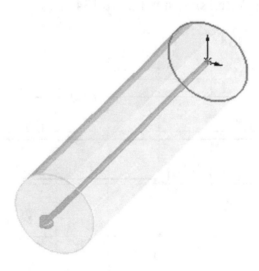

138. Your screen should look similar to Figure 133.

Figure 133

139. Save the part as Wristpin1.SLDPRT where it can be easily retrieved later.

140. Begin a new sketch as described in Chapter 1.

141. Complete the sketch shown in Figure 134.

Figure 134

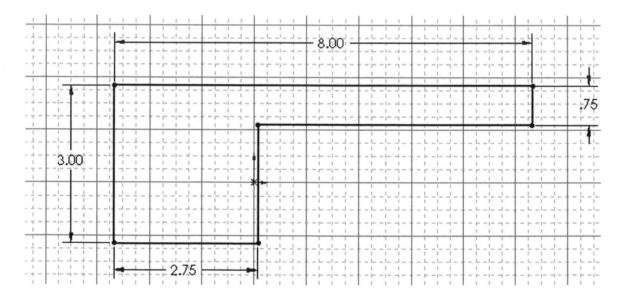

142. Exit the Sketch commands and change the view to Trimetric. A trimetric view will be displayed as shown in Figure 135.

Figure 135

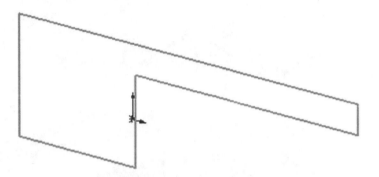

143. Extrude the sketch to a distance of **2.25** inches. Your screen should look similar to Figure 136.

Figure 136

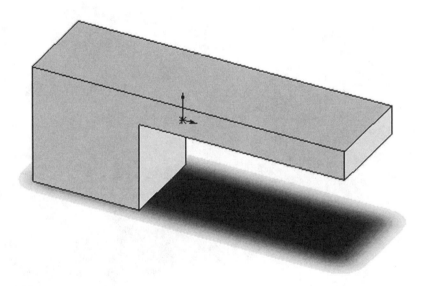

144. Use the **Fillet** command to create **1.125** inch fillets on the front portion of the part as shown in Figure 137.

Figure 137

Fillet Here

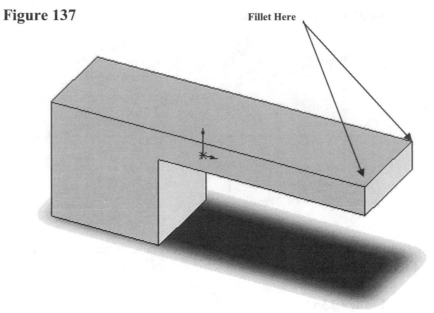

145. Your screen should look similar to Figure 138.

Figure 138

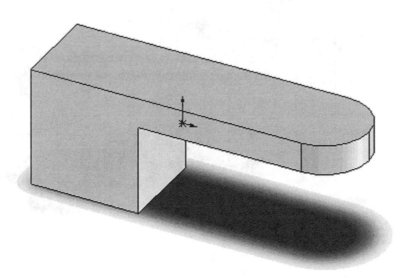

146. Move the cursor to the surface shown in Figure 139 causing it to turn red.

Figure 139

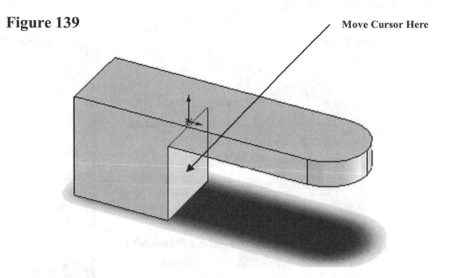

147. Left click once as shown in Figure 140. The surface will turn green.

Figure 140

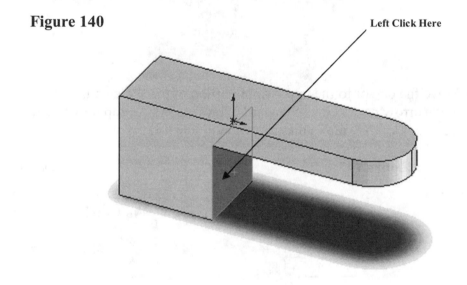

148. Right click on the surface. A pop up menu will appear. Left click on **Insert Sketch** as shown in Figure 141.

Figure 141

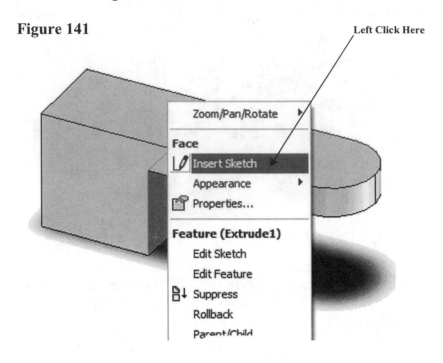

Left Click Here

149. Move the cursor to the upper right portion of the screen and left click on the drop down arrow next to the "Standard Views" icon. A drop down menu will appear. Left click on **Normal To** as shown in Figure 142.

Figure 142

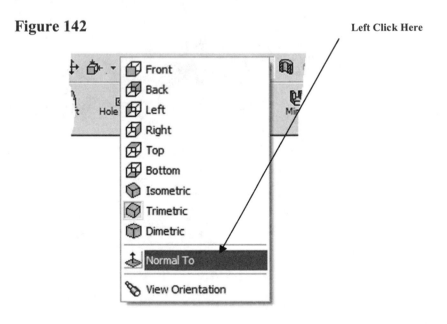

Left Click Here

150. A perpendicular view of the surface will be displayed as shown in Figure 143.

Figure 143

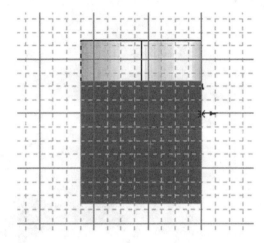

151. Create a sketch on the selected surface as shown in Figure 144.

Figure 144

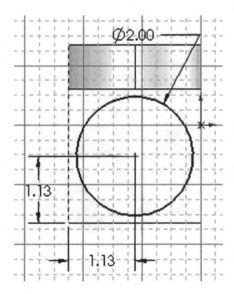

152. Exit out of the Sketch commands and change the view to trimetric as shown in Figure 145.

Figure 145

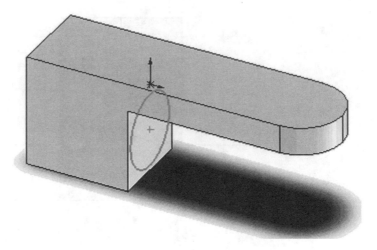

153. Use the **Extruded Cut** command to extrude or "cut" out the circle that was just completed. Your screen should look similar to Figure 146.

Figure 146

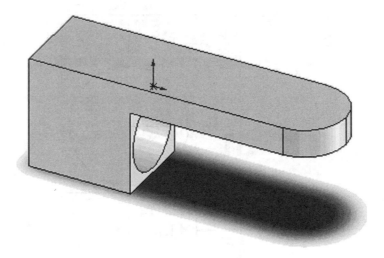

154. Left click on the top portion of the part causing it to turn green as shown in Figure 147.

Figure 147

Left Click Here

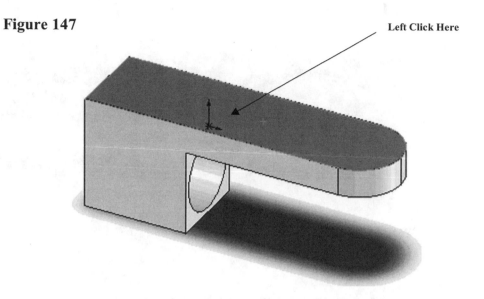

155. Move the cursor to the upper right portion of the screen and left click on the drop down arrow next to the "Standard Views" icon. A drop down menu will appear. Left click on **Normal To** as shown in Figure 148.

Figure 148

Left Click Here

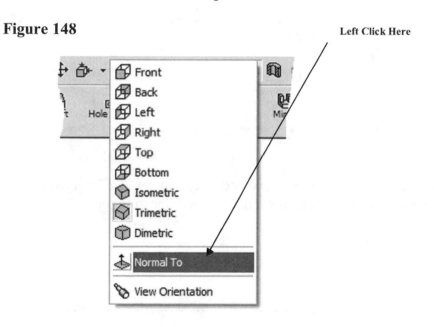

156. A perpendicular view will be displayed as shown in Figure 149.

Figure 149

157. Draw the sketch as shown in Figure 150.

Figure 150

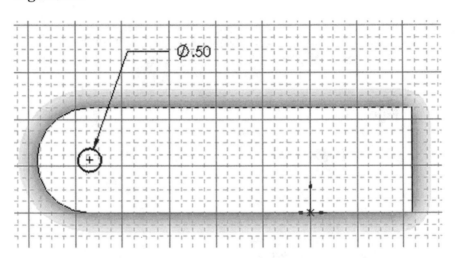

158. Use the **Extruded Cut** command to extrude or "cut" out the circle that was just completed. Your screen should look similar to Figure 151.

Figure 151

159. Change the view to trimetric. Your screen should look similar to Figure 152.

Figure 152

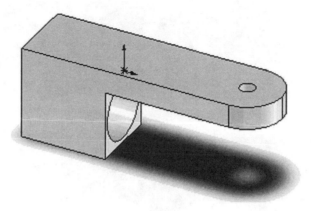

160. Save the part as Pistoncase1.ipt where it can be easily retrieved later.

161. Begin a new drawing as described in Chapter 1.

162. Complete the sketch shown in Figure 153.

Figure 153

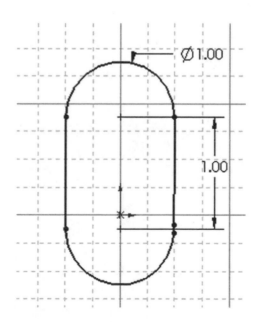

163. Extrude the sketch into a solid with a thickness of **.25** as shown in Figure 154.

Figure 154

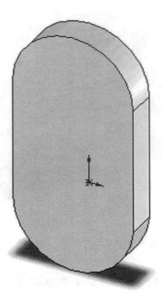

164. Complete the following sketch. Use the center of the outside fillet radius as the center of the circle as shown in Figure 155.

Figure 155

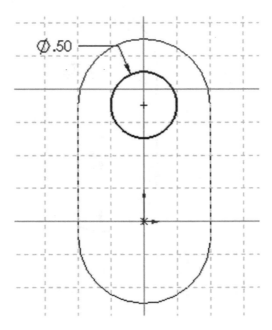

165. Extrude the sketch into a solid with a thickness of **.25** as shown in Figure 156.

Figure 156

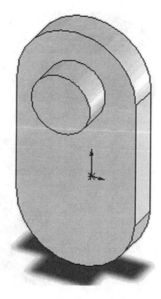

166. Use the rotate command and roll the part around to gain access to the opposite side as shown in Figure 157.

Figure 157

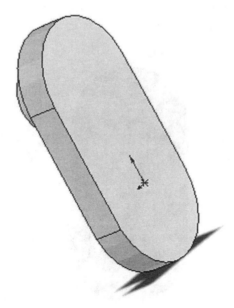

167. Begin a new sketch on the opposite side as shown in Figure 158.

Figure 158

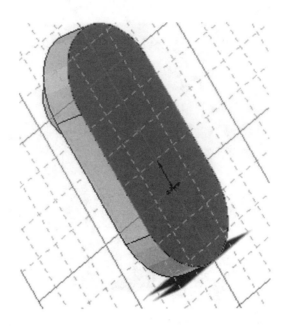

168. Use the **Normal To** command to gain a perpendicular view as shown in Figure 159.

Figure 159

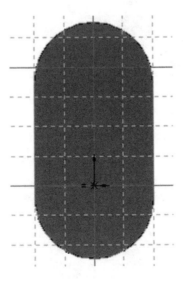

169. Complete the following sketch as shown in Figure 160.

Figure 160

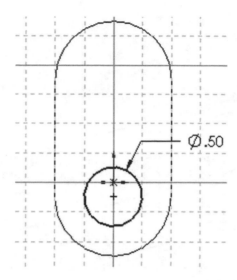

Ø.50

170. Extrude the sketch into a solid with a thickness of **.25** as shown in Figure 161.

Figure 161

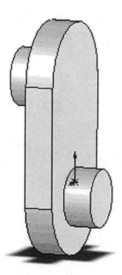

171. Save the part as Crankshaft1.ipt where it can be easily retrieved later.

172. Begin a new drawing as described in Chapter 1.

173. Complete the sketch shown in Figure 162. Extrude the sketch into a solid with a thickness of **.25**.

Figure 162

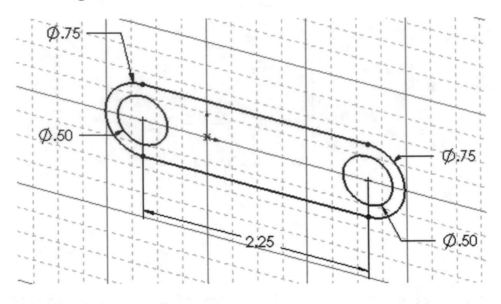

174. Save the part as Conrod1.ipt where it can be easily retrieved later.

175. All of these parts will be used in the next chapter.

Chapter 7 Introduction to Assembly
View Procedures

Objectives:

- Learn to import existing solid models into an Assembly
- Learn to constrain all parts in an Assembly
- Learn to assign colors to different parts in an Assembly
- Learn to edit/modify parts while in an Assembly
- Learn to create a rotary motor to simulate motion in an Assembly

Chapter 7 includes instruction on how to construct the assembly shown below.

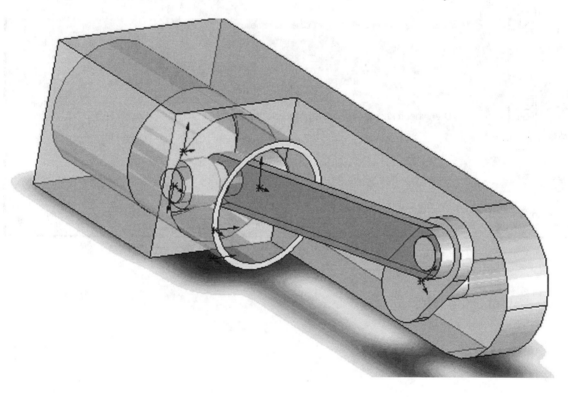

1.   Start SolidWorks by referring to "Chapter 1 Getting Started".

2.   After SolidWorks is running, begin an assembly drawing. Move the cursor to the upper left corner of the screen and left on **New**. The New SolidWorks Document dialog box will appear. Left click on **Assembly** as shown in Figure 1.

Figure 1   Left Click Here

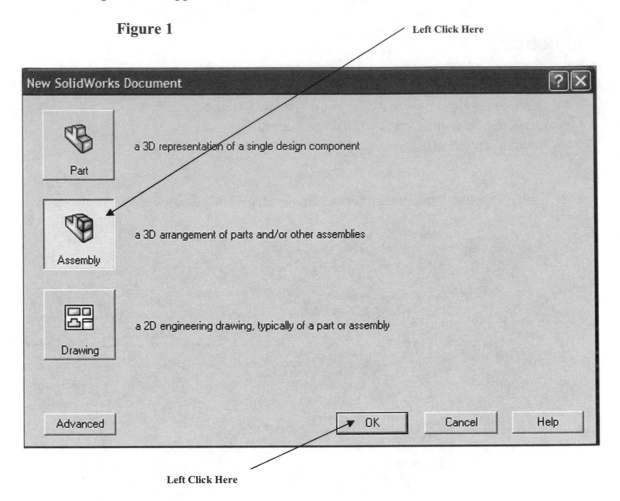

Left Click Here

3.   Left click on **OK**.

4. The Assembly feature of SolidWorks will open. Your screen should look similar to Figure 2.

Figure 2

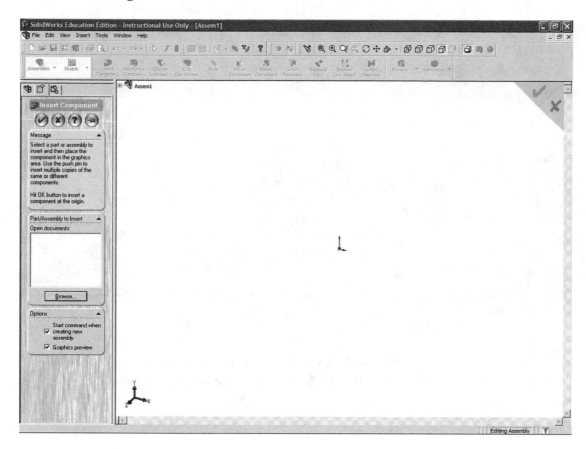

5. Move the cursor to the lower left portion of the screen and left click **Browse** as shown in Figure 3.

Figure 3

Left Click Here

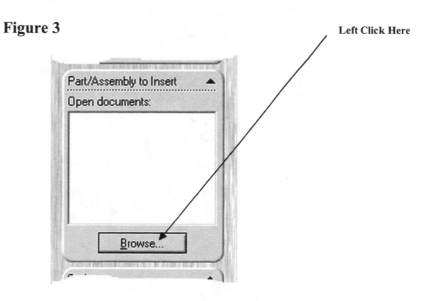

6. The Open dialog box will appear. Locate the Case1.SLDPART file and left click on **Open** as shown in Figure 4.

Figure 4

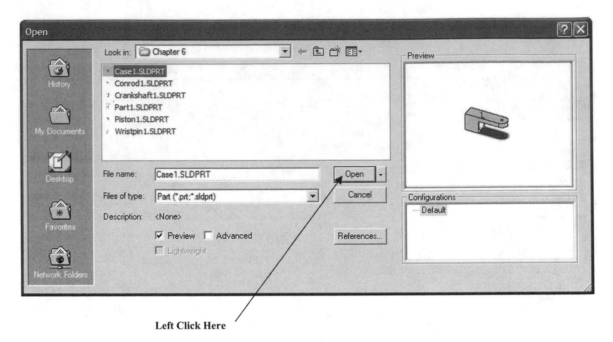

Left Click Here

7. SolidWorks will place one piston case in the drawing space. The case will be attached to the cursor. The piston case will appear as shown in Figure 5.

Figure 5

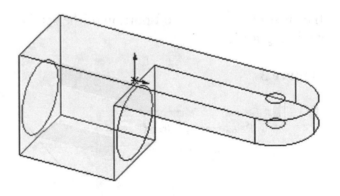

8. Left click once. The piston case will be placed in the assembly drawing. Your screen should look similar to Figure 6.

Figure 6

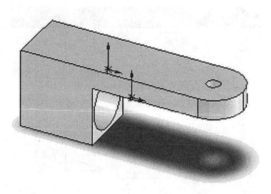

9. Move the cursor to the upper left portion of the screen and left click on **Insert Component** as shown in Figure 7.

Figure 7   Left Click Here

10. Move the cursor to the lower left portion of the screen and left click on **Browse** as shown in Figure 8.

Figure 8   Left Click Here

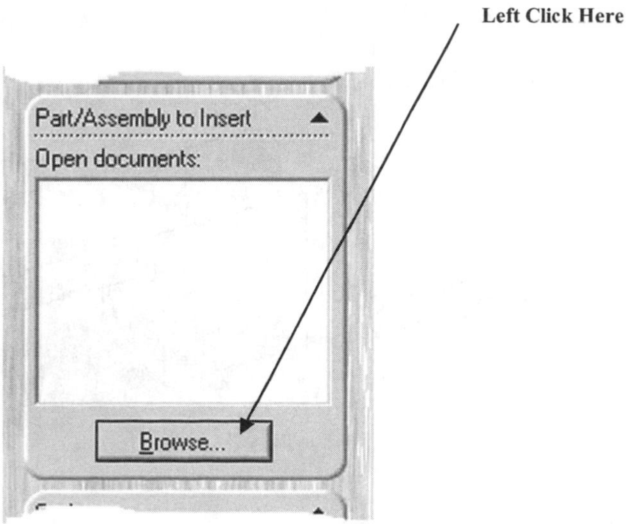

11.   The Open dialog box will appear. Locate the Piston1.SLDPRT file and left click on **Open** as shown in Figure 9.

Figure 9

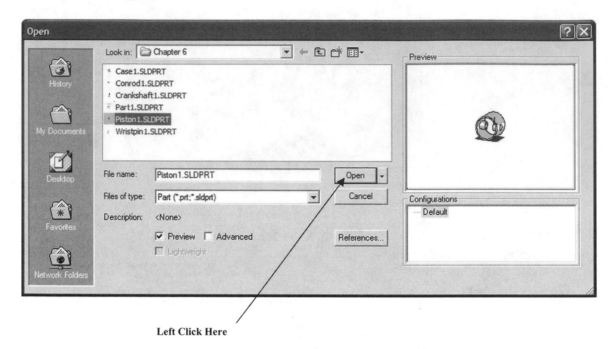

Left Click Here

12.   The piston will be attached to the cursor. Place the piston anywhere near the piston case and left click once. Your screen should look similar to Figure 10.

Figure 10

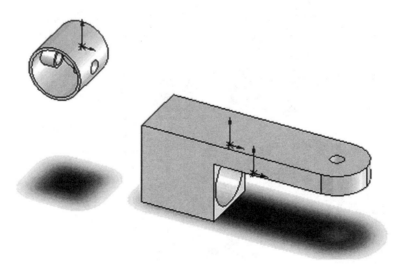

13. Move the cursor to the upper left portion of the screen and left click on **Insert Component** as shown in Figure 11.

Figure 11

Left Click Here

14. Move the cursor to the lower left portion of the screen and left click on **Browse** as shown in Figure 12.

Figure 12

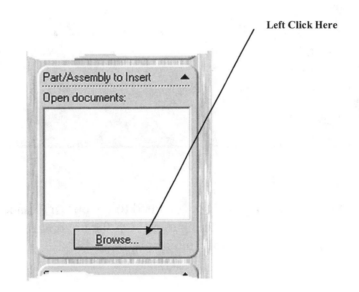

Left Click Here

15. The Open dialog box will appear. Locate the Conrod1.ipt file and left click on **Open** as shown in Figure 13.

Figure 13

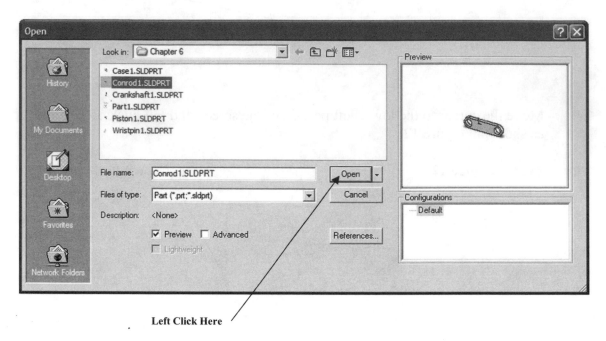

Left Click Here

16. The connecting rod will be attached to the cursor. Place the piston anywhere near the piston case and left click once. On the keyboard press the **Esc** button once. Your screen should look similar to Figure 14.

Figure 14

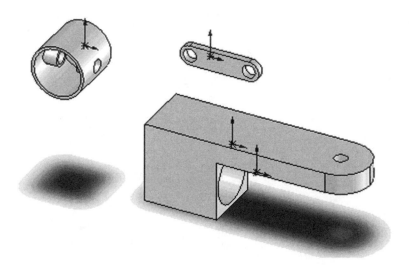

17. Move the cursor to the upper left portion of the screen and left click on **Insert Component** as shown in Figure 15.

Figure 15   Left Click Here

18. Move the cursor to the lower left portion of the screen and left click on **Browse** as shown in Figure 16.

Figure 16   Left Click Here

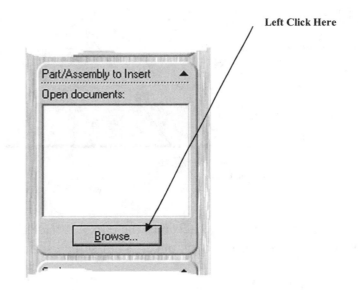

19. The Open dialog box will appear. Locate the crankshaft1.SLDPRT file and left click on **Open** as shown in Figure 17.

Figure 17

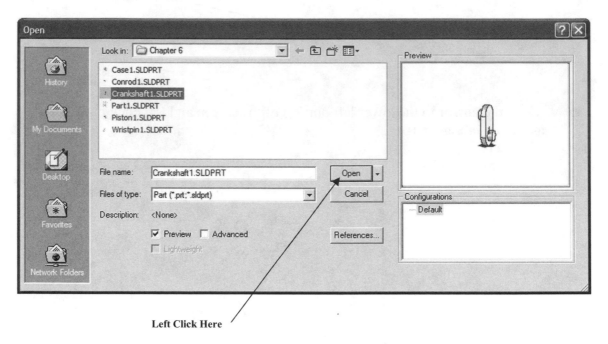

Left Click Here

20. The crankshaft will be attached to the cursor. Place the crankshaft anywhere near the piston case and left click once. Your screen should look similar to Figure 18.

Figure 18

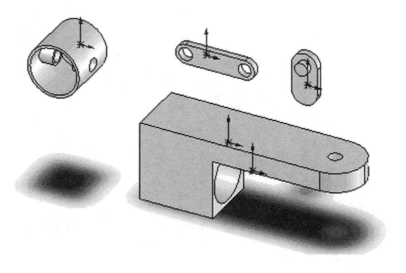

21. Move the cursor to the upper left portion of the screen and left click on **Insert Component** as shown in Figure 19.

Figure 19

Left Click Here

22. Move the cursor to the lower left portion of the screen and left click on **Browse** as shown in Figure 20.

Figure 20

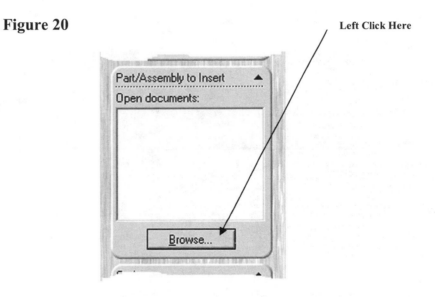

Left Click Here

23.   The Open dialog box will appear. Locate the wristpin1.SLDPRT file and left click on **Open** as shown in Figure 21.

Figure 21

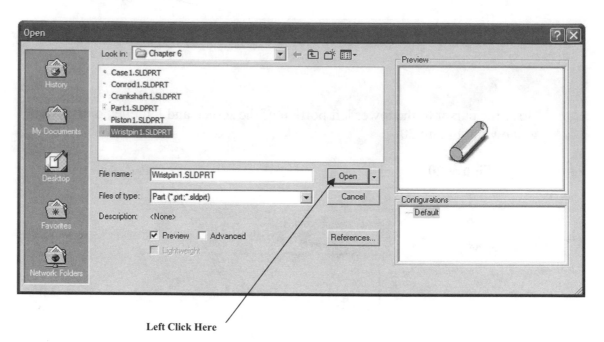

Left Click Here

24.   The wristpin will be attached to the cursor. Place the wristpin anywhere near the piston case and left click once. Your screen should look similar to Figure 22.

Figure 22

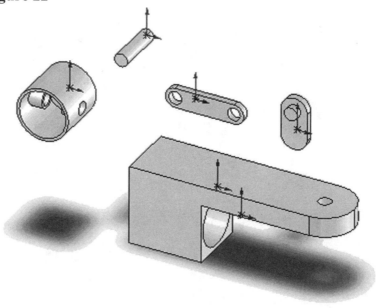

25.   The first part inserted into the assembly becomes "fixed" meaning it cannot be moved. All other parts can be moved by left clicking (holding the left mouse button down) and dragging the part to the desired location.

26.   Use the Rotate command to rotate the entire parts group to the position shown in Figure 23.

Figure 23

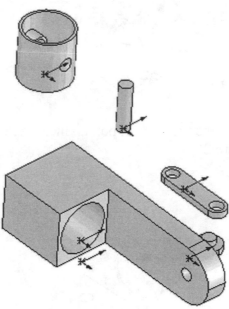

27.   Move the cursor to the upper middle portion of the screen and left click on **Rotate Component** as shown in Figure 24.

Figure 24   Left Click Here

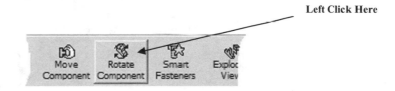

28. Left click (holding the left mouse button down) and rotate the piston to a horizontal position as shown in Figure 25.

Figure 25

29. Right click anywhere around the part. A pop up menu will appear. Left click on **Select** as shown in Figure 26.

Figure 26

Left Click Here

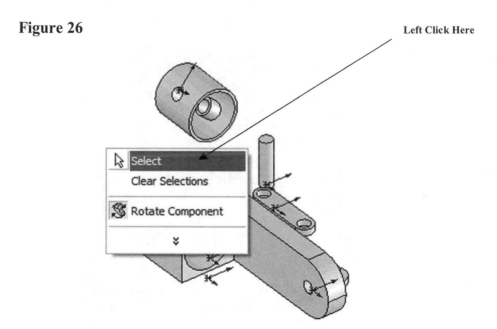

30. Move the cursor to the upper middle portion of the screen and left click on **Mate** as shown in Figure 27.

Figure 27

Left Click Here

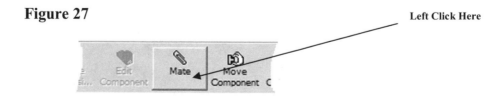

340

31.   Move the cursor to the piston causing the OUTSIDE edges of the part to turn red. Left click once. The part will turn green as shown in Figure 28.

Figure 28   Turned Green

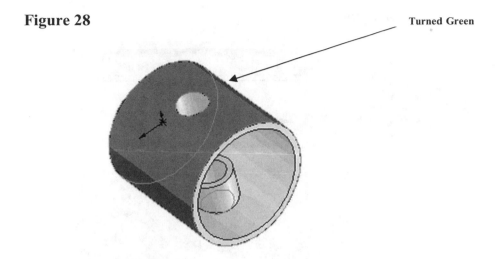

32.   Move the cursor to the piston case causing the cylinder edges to turn red. Left click once. The cylinder will turn green as shown in Figure 29.

Figure 29   Turned Green

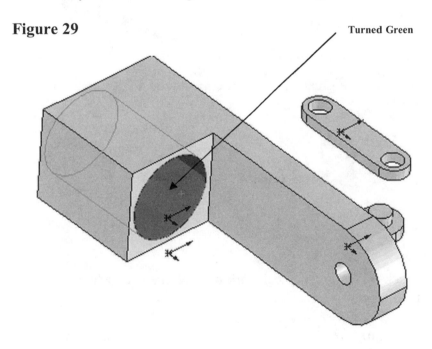

33. The centers of the piston and the cylinder will be aligned. Your screen should look similar to Figure 30.

Figure 30

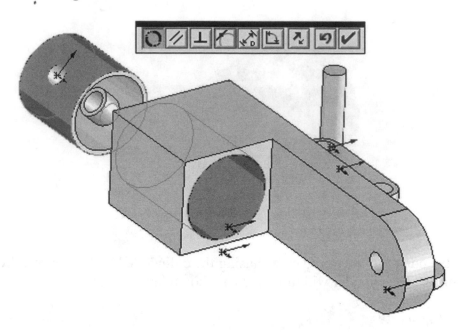

34. Left click on the green checkmark beneath the text "Concentric1"as shown in Figure 31.

Figure 31   Left Click Here

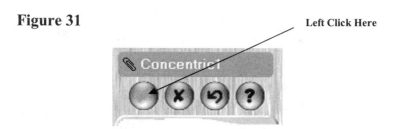

35. Left click on the green checkmark beneath the text "Mate" as shown in Figure 32.

Figure 32   Left Click Here

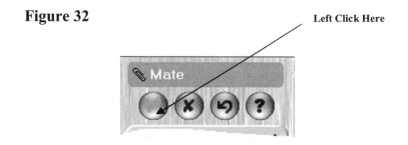

36. Left click (holding the left mouse button down) and drag the piston down into the cylinder as shown in Figure 33.

Figure 33

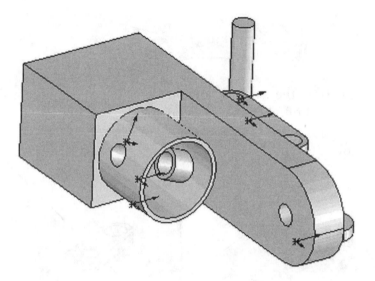

37. Move the cursor to the lower left portion of the piston. Left click (holding the left mouse button down) and slide the piston downward, out below the bore as shown in Figure 34.

Figure 34

38.   Move the cursor to the upper middle portion of the screen and left click on **Mate** as shown in Figure 35.

Figure 35   Left Click Here

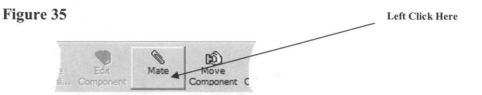

39.   Move the cursor to the wristpin hole on the piston causing the inside edge of the hole to turn red. Left click once as shown in Figure 36.

Figure 36   Left Click Here

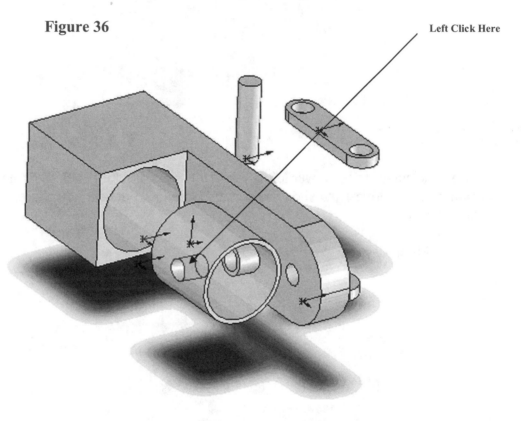

40. Move the cursor to the upper portion of the connecting rod causing the inside edges of the hole to turn red. You may have to drag the connecting rod to a position where the end holes are accessible. Left click as shown in Figure 37.

Figure 37

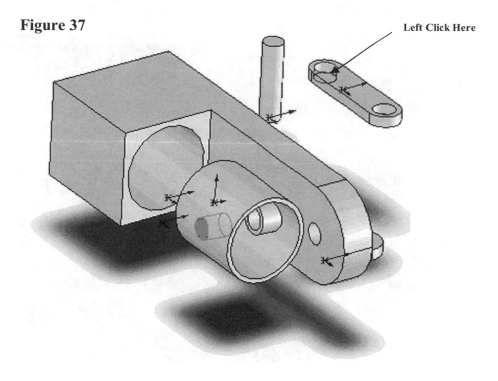

Left Click Here

41.   The connecting rod hole and the center of the wristpin hole in the piston will be
        aligned as shown in Figure 38.

        Figure 38

42.   Left click on the green checkmark beneath the text "Concentric2" as shown in
        Figure 39.

        Figure 39   **Left Click Here**

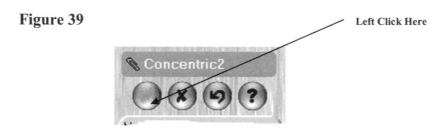

43.   Left click on the green checkmark beneath the text "Mate" as shown in Figure 40.

        Figure 40   **Left Click Here**

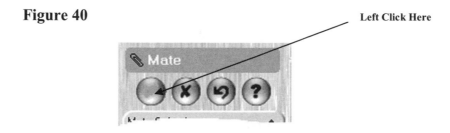

44. Use the rotate command to rotate the entire assembly around to gain access to the underside of the piston. Move the connecting rod to the location shown in Figure 41. This will take some skill. If the piston moves during this process, use the cursor to move the piston to the original location.

Figure 41

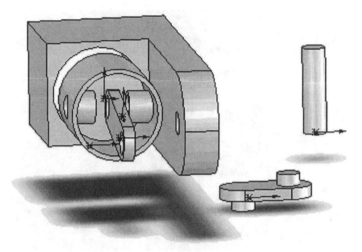

45. Move the cursor to the upper middle portion of the screen and left click on **Mate** as shown in Figure 42.

Figure 42   Left Click Here

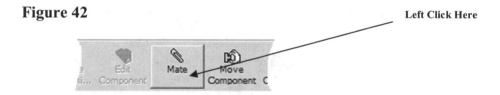

46. Move the cursor to the left side of the connecting rod causing the edges of the connecting rod to turn red. Left click as shown in Figure 43. You may have to zoom in so that SolidWorks will find the proper surface.

Figure 43   Left Click Here

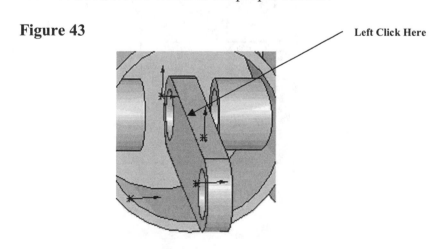

47.   Use the rotate command to move the piston to gain access to the surface opposite the previously selected surface as shown in Figure 44. After the assembly is rotated, press the **Esc** key on the keyboard to get out of the rotate command. Select the second mating surface by left clicking as shown in Figure 44.

Figure 44

Left Click Here

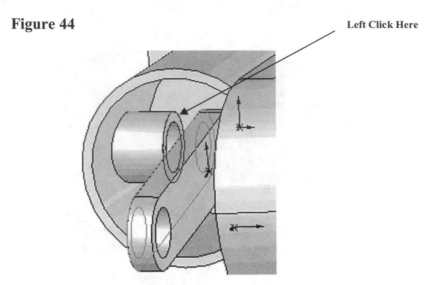

48.   Left click on **D** and enter **.250** as shown in Figure 45.

Figure 45

Left Click Here   Enter .250 Here

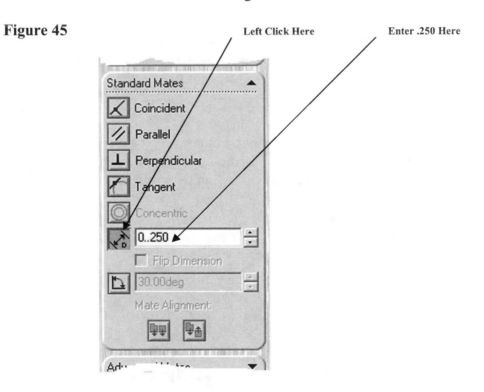

49.   Left click on the green checkmark beneath the text "Distance1" as shown in Figure 46.

Figure 46

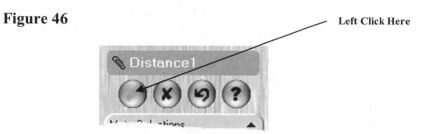

50.   Left click on the green checkmark beneath the text "Mate" as shown in Figure 47.

Figure 47

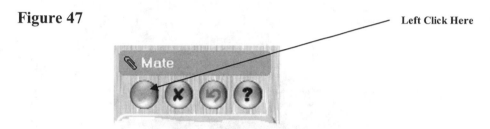

51.   The connecting rod should be centered in the piston. Your screen should look similar to Figure 48.

Figure 48

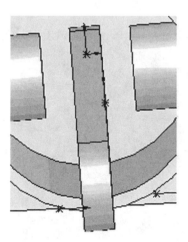

52.   Use the rotate command to rotate the entire assembly to the position shown in Figure 49.

Figure 49

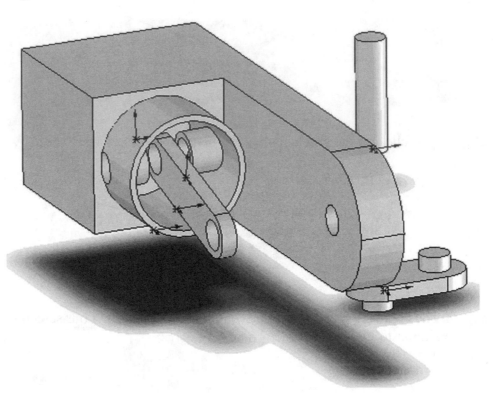

53.   Move the cursor to the upper middle portion of the screen and left click on **Mate** as shown in Figure 50.

Figure 50   Left Click Here

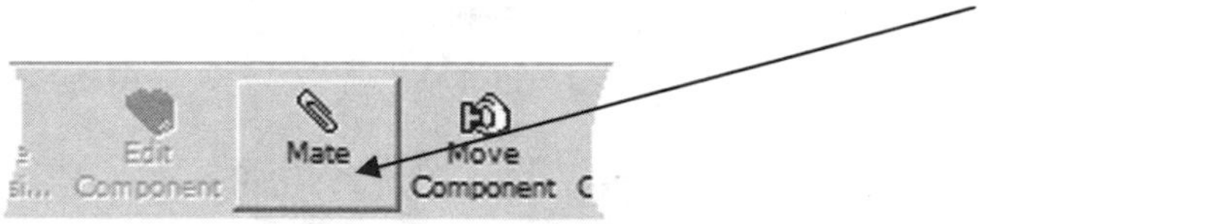

54. Move the cursor to the wristpin causing the edges of the wristpin to turn red. Left click once as shown in Figure 51.

Figure 51

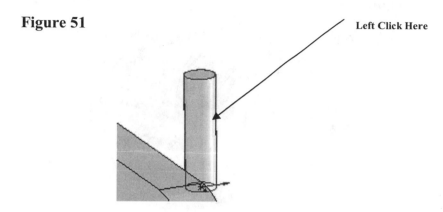

Left Click Here

55. Move the cursor to the wristpin hole causing the edges of the hole to turn red. Left click once as shown in Figure 52.

Figure 52

Left Click Here

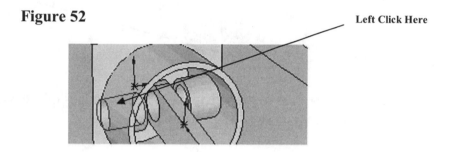

56. Left click on the green checkmark beneath the text "Concentric3" as shown in Figure 53.

Figure 53

Left Click Here

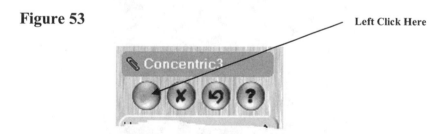

57. Left click on the green checkmark beneath the text "Mate" as shown in Figure 54.

Figure 54

Left Click Here

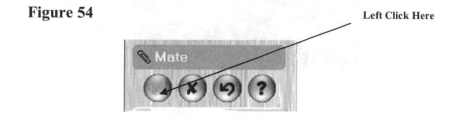

58.   The wristpin will be placed in the wristpin hole as shown in Figure 55.

Figure 55

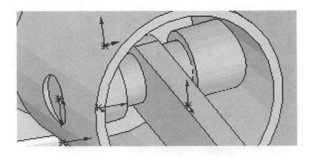

59.   Move the cursor to the upper middle portion of the screen and left click on **Mate** as shown in Figure 56.

Figure 56   Left Click Here

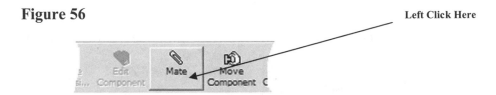

60.   Move the cursor to the side of the wristpin causing the edges of the wristpin to turn red. Left click once as shown in Figure 57.

Figure 57   Left Click Here

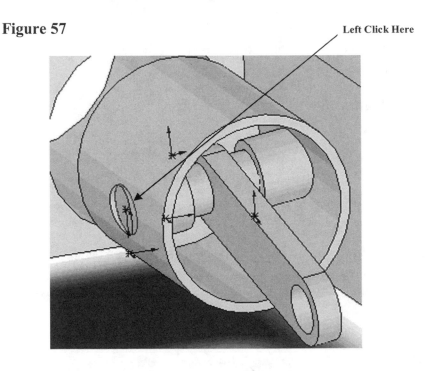

61. Move the cursor to the side of the connecting rod causing the edges of the connecting rod to turn red. Left click once as shown in Figure 58.

Figure 58

Left Click Here

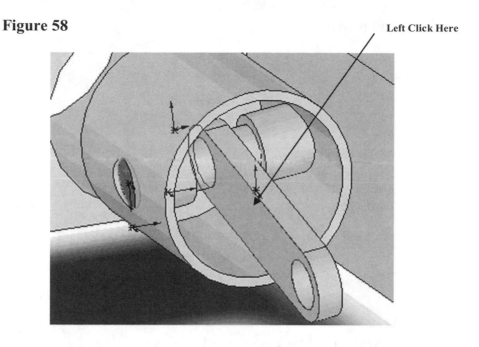

62. Left click on **D** and enter **.7825** as shown in Figure 59.

Figure 59

Left Click Here   Enter .7825 Here

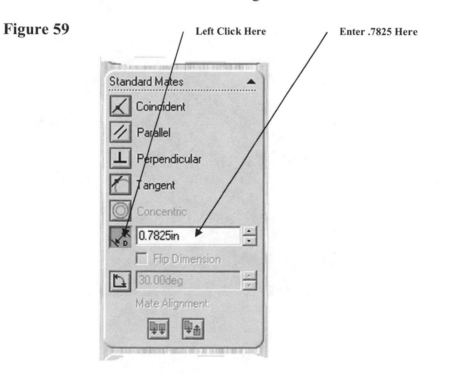

63. Left click on the green checkmark beneath the text "Distance2" as shown in Figure 60.

Figure 60   Left Click Here

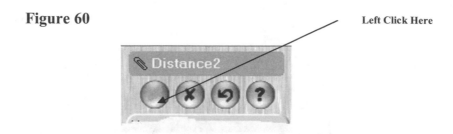

64. Left click on the green checkmark beneath the text "Mate" as shown in Figure 61.

Figure 61   Left Click Here

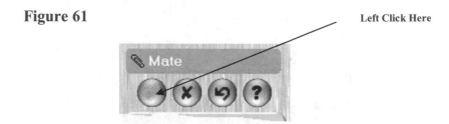

65. Your screen should look similar to Figure 62.

Figure 62

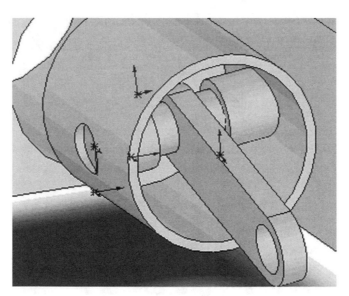

66. Move the cursor to the upper middle portion of the screen and left click on **Move Component** as shown in Figure 63.

Figure 63

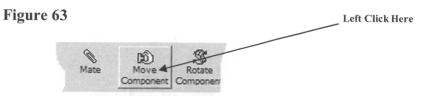

Left Click Here

67. Left click again (holding the left mouse button down) and drag the crankshaft to where it can be rotated as shown in Figure 64.

Figure 64

Left Click Here

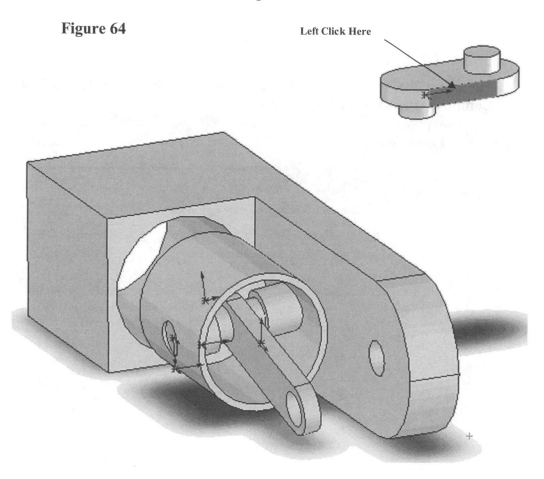

68. Right click anywhere around the parts. A pop up menu will appear. Left click on **Select** as shown in Figure 65.

Figure 65

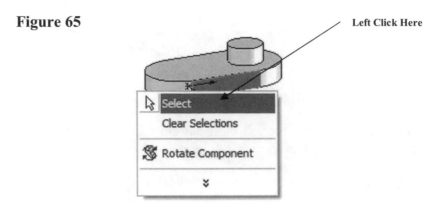

Left Click Here

69. Move the cursor to the upper middle portion of the screen and left click on **Rotate Component** as shown in Figure 66.

Figure 66

Left Click Here

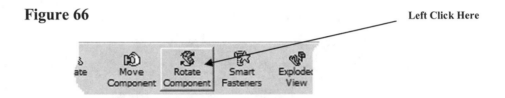

70. Left click once. Left click again (holding the left mouse button down) and rotate the connecting rod upward as shown in Figure 67.

Figure 67

Rotate Upward

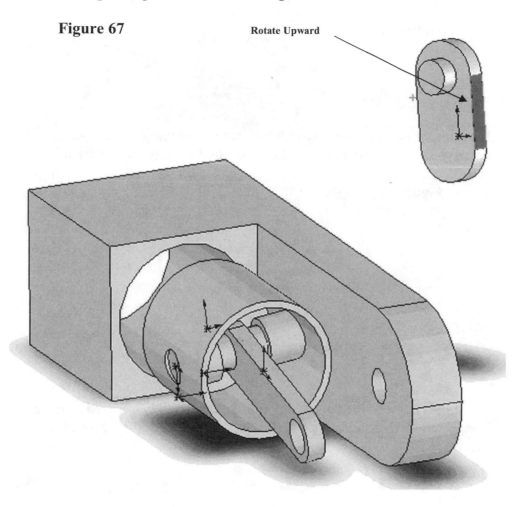

71. Right click anywhere around the parts. A pop up menu will appear. Left click on **Select** as shown in Figure 68.

Figure 68

Left Click Here

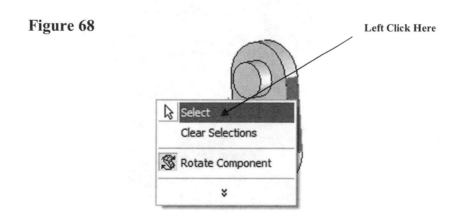

72.　Move the cursor to the upper middle portion of the screen and left click on **Mate** as shown in Figure 69.

Figure 69

Left Click Here

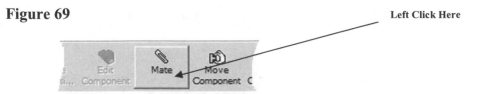

73.　Move the cursor to the crankshaft pin that will secure the connecting rod. The edges of the crankshaft pin will turn red. Left click once as shown in Figure 70.

Figure 70

Left Click Here

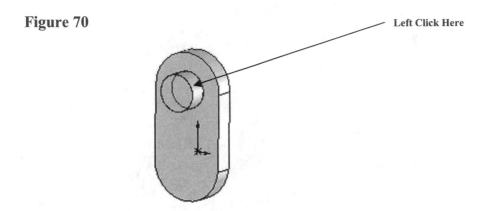

74.　Move the cursor to the connecting rod hole that will be secured to the crankshaft. The edges of the connecting rod hole will turn red. Left click once as shown in Figure 71.

Figure 71

Left Click Here

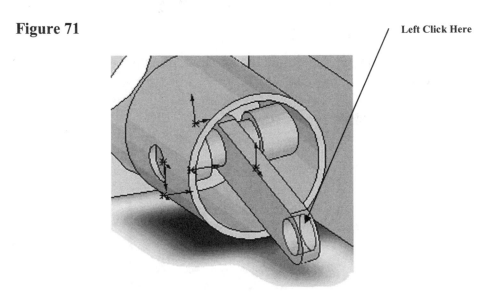

75. Left click on the green checkmark beneath the text "Concentric4" as shown in Figure 72.

Figure 72

Left Click Here

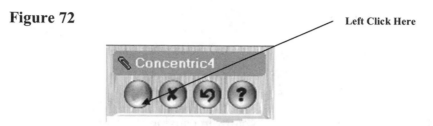

76. Left click on the green checkmark beneath the text "Mate" as shown in Figure 73.

Figure 73

Left Click Here

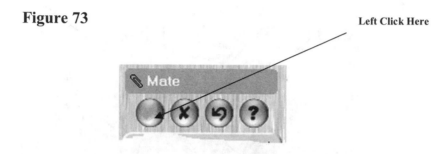

77. The center of the crankshaft and the center of the connecting rod are now the same.

78. Left click on the crankshaft (holding the left mouse button down) and drag it to the connecting rod as shown in Figure 74.

Figure 74

Drag Parts Together

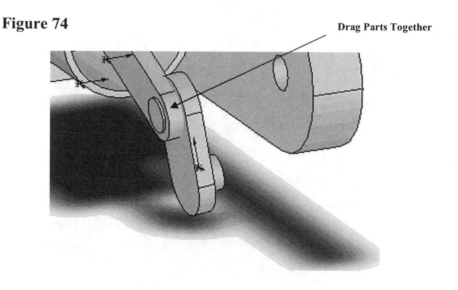

79. Move the cursor to the upper middle portion of the screen and left click on **Mate** as shown in Figure 75.

Figure 75

Left Click Here

80. Move the cursor to the crankshaft pin causing the edge of the pin to turn red. Left click as shown in Figure 76.

Figure 76

Left Click Here

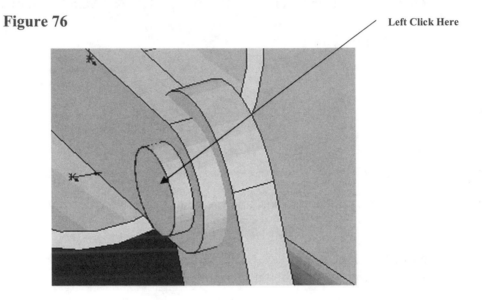

81.   Move the cursor to the connecting rod side. The edge of the connecting rod will turn red. Left click as shown in Figure 77.

Figure 77

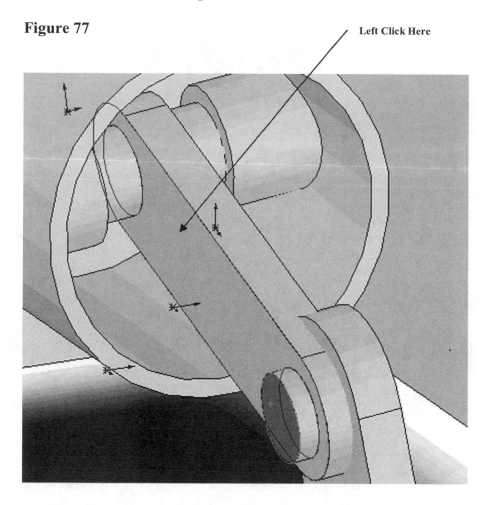

82.   Left click on the green checkmark beneath the text "Coincident5" as shown in Figure 78.

Figure 78

83. Left click on the green checkmark beneath thc text "Mate" as shown in Figure 79.

Figure 79

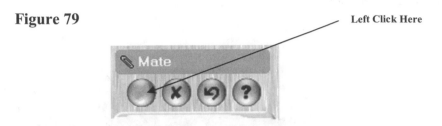

Left Click Here

84. SolidWorks will place the connecting rod and crankshaft together as shown in Figure 80.

Figure 80

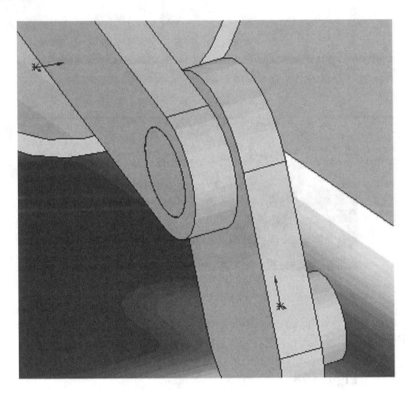

85. Use the rotate command and roll the assembly around to gain access to the opposite side as shown in Figure 81.

Figure 81

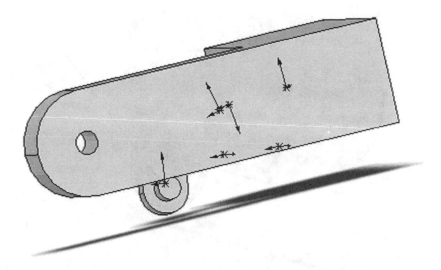

86. Move the cursor to the upper middle portion of the screen and left click on **Mate** as shown in Figure 82.

Figure 82

Left Click Here

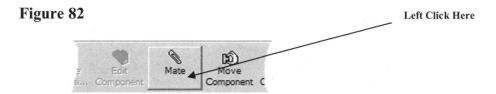

87. Move the cursor to the crankshaft pin that will be secured in the piston case. The edges of the pin will turn red. Left click once as shown in Figure 83.

Figure 83

Left Click Here

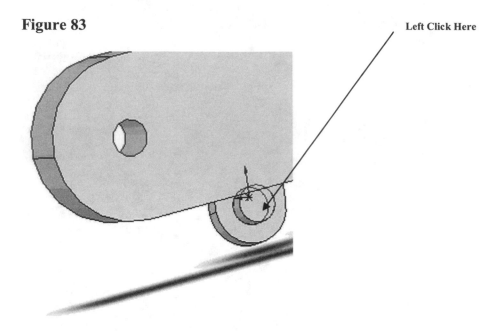

88. Move the cursor to the piston case hole that will secure the crankshaft. The edges of the hole will turn red. Left click once as shown in Figure 84.

Figure 84

Left Click Here

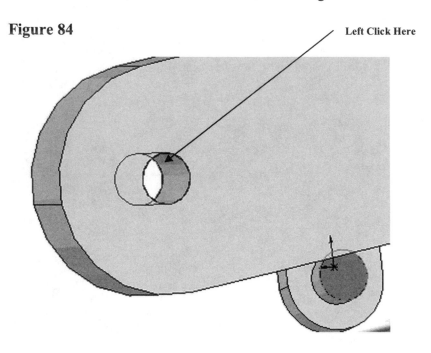

89. SolidWorks will place the crankshaft pin into the piston case as shown in Figure 85.

Figure 85

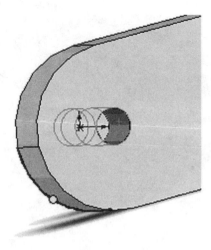

90. Left click on the green checkmark beneath the text "Concentric6" as shown in Figure 86.

Figure 86   Left Click Here

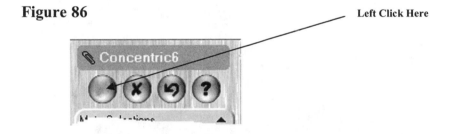

91. Left click on the green checkmark beneath the text "Mate" as shown in Figure 87.

Figure 87   Left Click Here

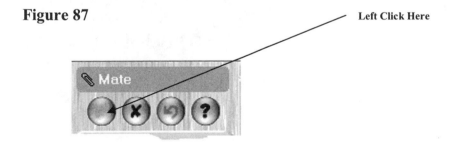

92.   Your screen should look similar to Figure 88.

Figure 88

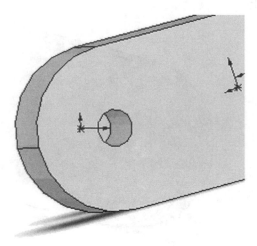

93.   Use the **Rotate** command to roll the parts around to view the assembly. Your screen should look similar to Figure 89.

Figure 89

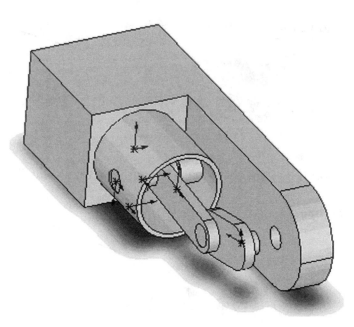

94. The length of the connecting rod must be modified. Move the cursor over the connecting rod as shown in Figure 90. The edges will turn red.

Figure 90

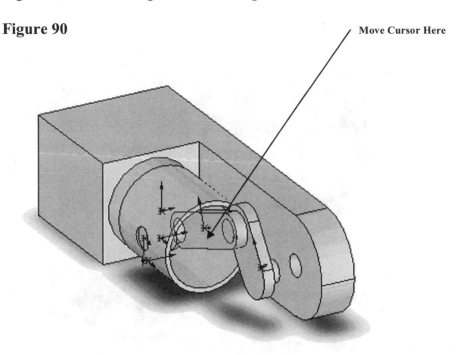

95. Left click once. The surface of the connecting rod will turn green. Right click once. A pop up menu will appear. Left click on **Edit Sketch** as shown in Figure 91.

Figure 91

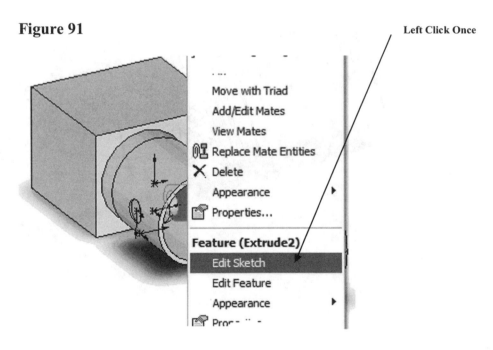

96.   The sketch used to create the connecting rod will be displayed as shown in Figure 92. Use the Rotate command to obtain a better view of the connecting rod.

Figure 92

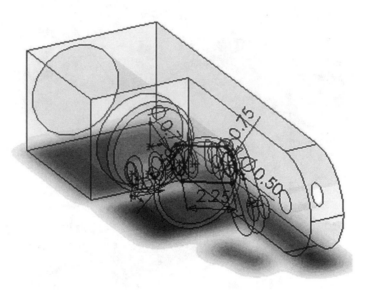

97.   Your screen should look similar to Figure 93.

Figure 93

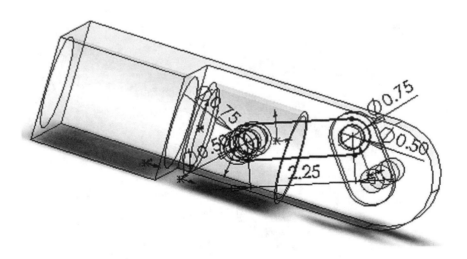

98.   Move the cursor over the 2.25 dimension. A red box will appear around the text. Double click. The Modify dialog box will appear as shown in Figure 94.

Figure 94

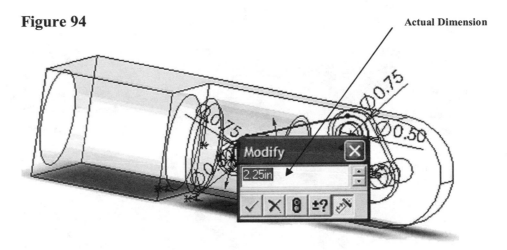

99.   While the text is still highlighted, enter **4.5** and left click on the green checkmark as shown in Figure 95.

Figure 95

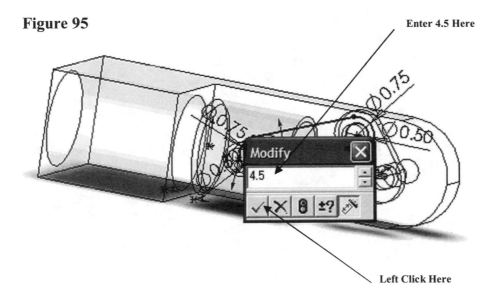

369

100. The length of the connecting rod will become 4.5 inches as shown in Figure 96.

Figure 96

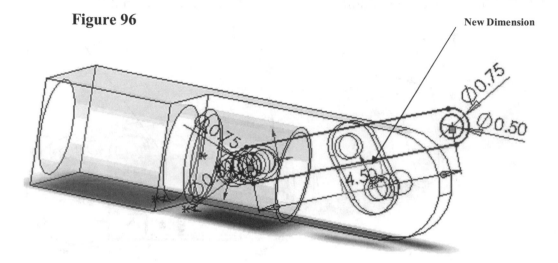

101. Left click on the green checkmark beneath the text "Dimension" as shown as shown in Figure 97.

Figure 97

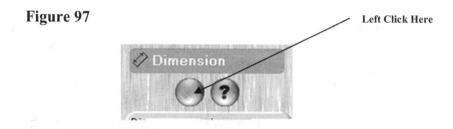

102. Right click anywhere around the parts. A pop up menu will appear. Left click on **Exit Sketch** as shown in Figure 98.

Figure 98

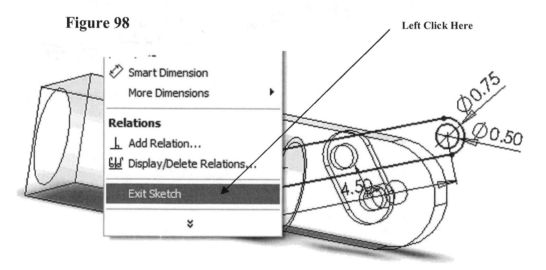

103. Right click anywhere around the parts. A pop up menu will appear. Left click on **Edit Assembly: Assem1** as shown in Figure 99.

Figure 99

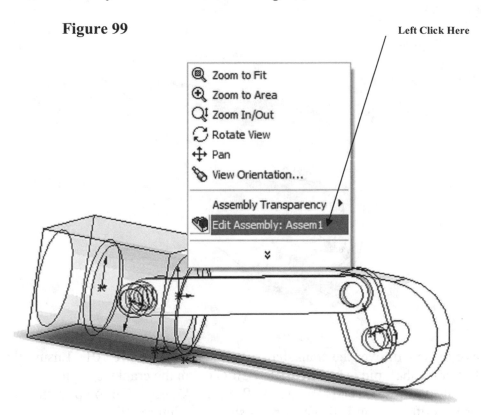

104. The changes made to the connecting rod will be displayed. Your screen should look similar to Figure 100.

Figure 100

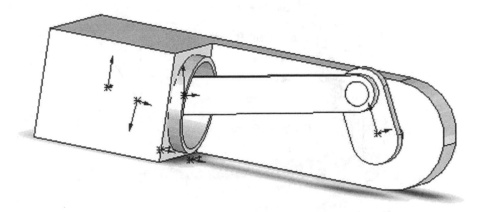

105. The length of the crankshaft pin also must be modified. Use the Rotate command to rotate the assembly to gain access to the backside of the crankshaft as shown in Figure 101.

Figure 101

Move Cursor Here

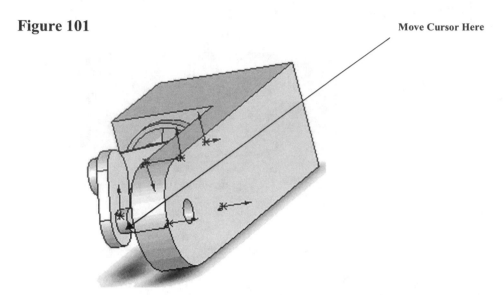

106. Move the cursor over the crankshaft pin as shown in Figure 101. Ensure that the entire crankshaft pin is highlighted. Left click on the crankshaft pin. The entire pin of the crankshaft will turn green. Right click once. A pop up menu will appear. Left click on **Edit Feature** as shown in Figure 102.

Figure 102

Left Click Here

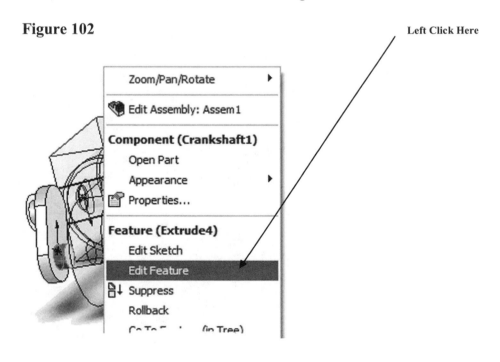

107. A preview of the extrusion will be displayed as shown in Figure 103.

Figure 103

Preview of 2.00 Extrusion

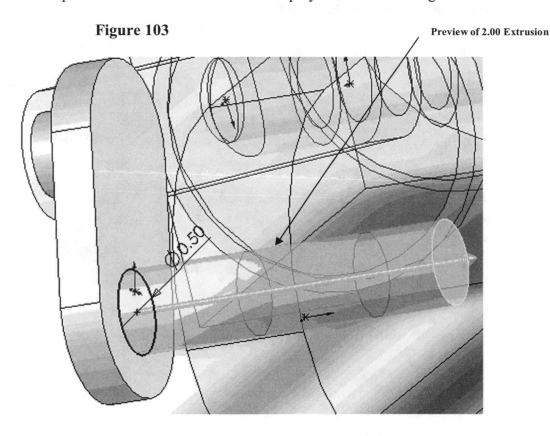

108. Enter **2.00** for the extrusion distance and left click on the green checkmark as shown in Figure 104.

Figure 104

Enter 2.00 Here   Left Click Here

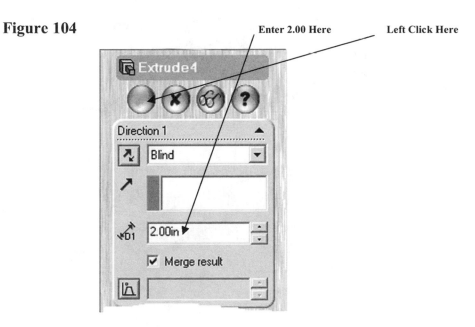

109. Right click anywhere around the parts. A pop up menu will appear. Left click on **Edit Assembly: Assem1** as shown in Figure 105.

Figure 105

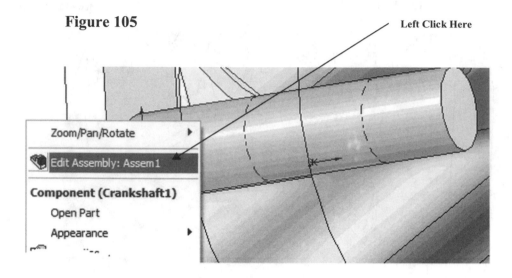

110. The changes made to the crankshaft will be displayed. Your screen should look similar to Figure 106.

Figure 106

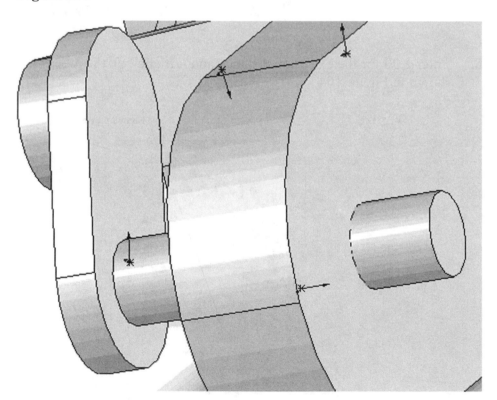

111. Use the Rotate command to position the part as shown in Figure 107.

Figure 107

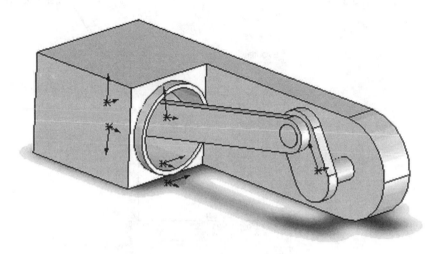

112. Move the cursor to the upper left portion of the screen and left click on the "Edit Color" icon as shown in Figure 108.

Figure 108   Left Click Here

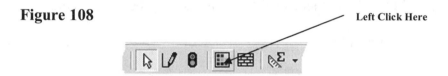

113. Move the cursor to the color palate and left click on light green as shown in Figure 109.

Figure 109   Left Click Here

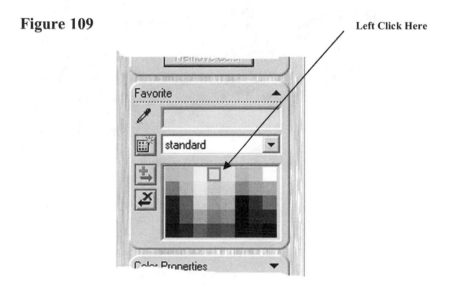

114. Left click on the drop down arrow next to the "Standard" text. A drop down menu will appear. Left click on **transparent** as shown in Figure 110.

Figure 110

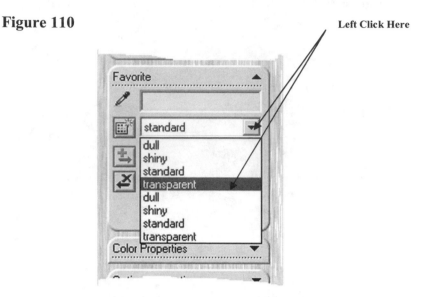

115. Move the cursor to the left side of the piston case and left click once as shown in Figure 111. The left side of the piston case will become green.

Figure 111

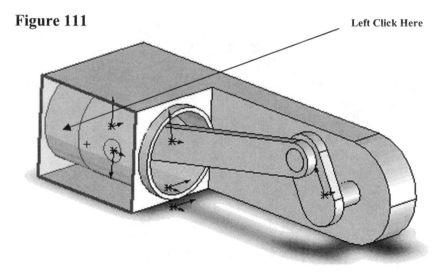

116. Left click on the top surface of the piston case, the underside where the bottom of the bore is, and the inside surface of the piston case as shown in Figure 112.

Figure 112

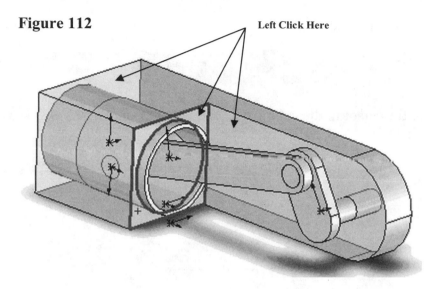

Left Click Here

117. Left click on the green checkmark beneath the text "Color And Optics" as shown in Figure 113.

Figure 113

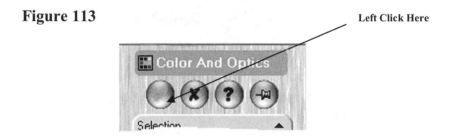

Left Click Here

118. The color of the piston case will change to clear green as shown in Figure 114.

Figure 114

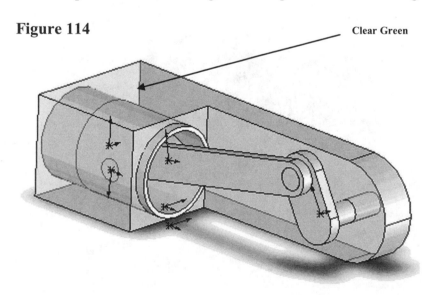

Clear Green

119. Move the cursor to the upper left portion of the screen and left click on the "Edit Color" icon as shown in Figure 115.

Figure 115

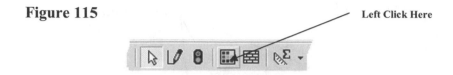

Left Click Here

120. Move the cursor to the color palate and left click on light blue as shown in Figure 116.

Figure 116

Left Click Here

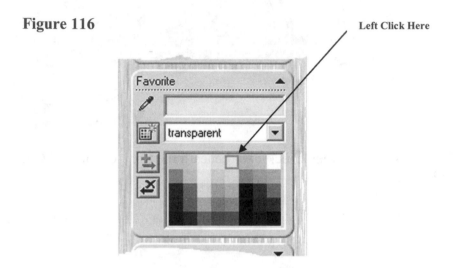

121. Move the cursor to any portion of the piston. The edges will turn red. Left click once as shown in Figure 117.

Figure 117

Left Click Here

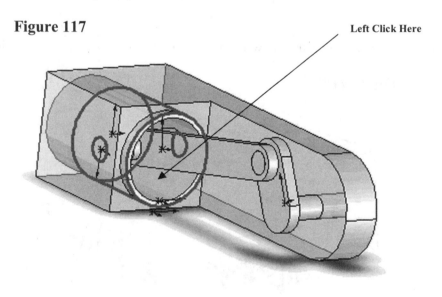

122. Left click on the green checkmark beneath the text "Color And Optics" as shown in Figure 118.

Figure 118

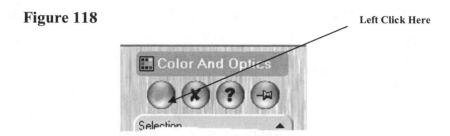

Left Click Here

123. The color of the piston will change to clear polished blue as shown in Figure 119.

Figure 119

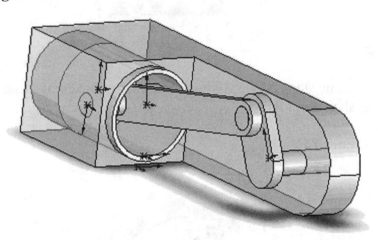

124. Move the cursor to the upper left portion of the screen and left click on the "Edit Color" icon as shown in Figure 120.

Figure 120

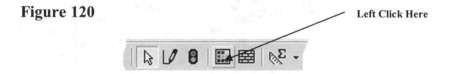

Left Click Here

125. Move the cursor to the color palate and left click on light brown as shown in Figure 121.

Figure 121

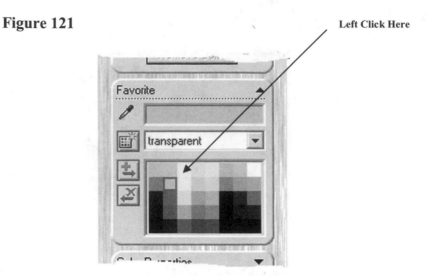

126. Using the same steps previously outlined, change the connecting rod color to light brown as shown in Figure 122.

Figure 122

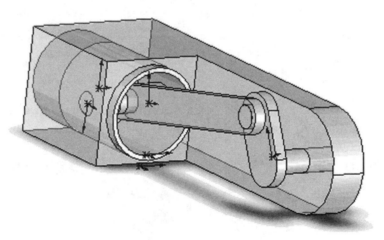

127. Left click on the green checkmark beneath the text "Color And Optics" as shown in Figure 123.

Figure 123

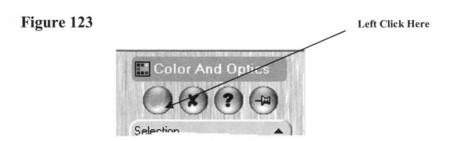

128. Move the cursor to the upper right portion of the screen and left click on the drop down arrow next to the "Simulation" text. A drop down box will appear. Left click on **Rotary Motor** as shown in Figure 124.

Figure 124

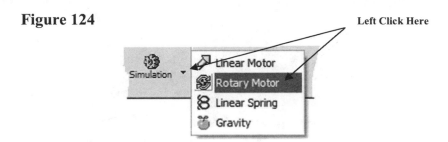

Left Click Here

129. Move the cursor to the crankshaft pin that protrudes through the piston case. The crankshaft pin will turn red as shown in Figure 125.

Figure 125

Turned Red

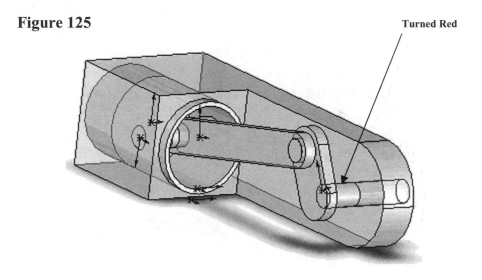

130. Left click once. A red arrow will be displayed showing the direction of motion as shown in Figure 126.

Figure 126

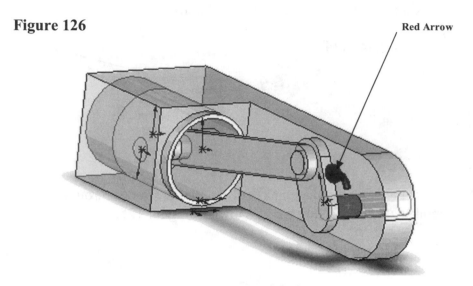

131. Left click on the green checkmark beneath the text "Rotary Motor" as shown in Figure 127.

Figure 127

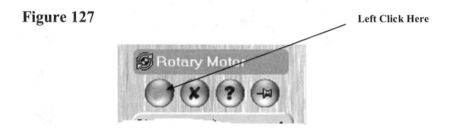

132. Move the cursor to the upper right portion of the screen and left click on the drop down arrow next to the "Simulation" text. A drop down menu will appear. Left click on **Calculate Simulation** as shown in Figure128.

Figure 128

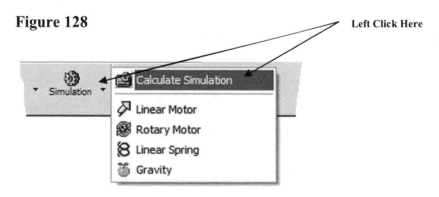

133. SolidWorks will begin animating the part. If SolidWorks detects any part collision or interference, the animation will stop playing. To stop the animation, press the **Esc** key on the keyboard.

134. Save the part where it can easily be retrieved. When closing SolidWorks, a dialog box will appear indicating that models making up the assembly have been modified. Left click on **Yes** as shown in Figure 129.

Figure 129

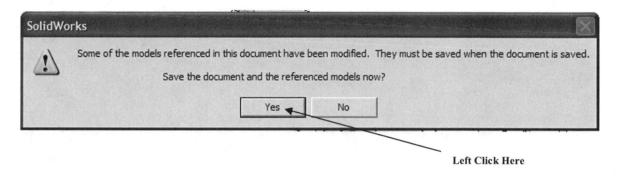

Left Click Here

INDEX